AF743616

CHINA IN THE 21ST CENTURY

POLLUTION IN CHINA

CHINA IN THE 21ST CENTURY

Additional books in this series can be found on Nova's website under the Series tab.

Additional E-books in this series can be found on Nova's website under the E-books tab.

ENVIRONMENTAL SCIENCE, ENGINEERING, AND TECHNOLOGY

Additional books in this series can be found on Nova's website under the Series tab.

Additional E-books in this series can be found on Nova's website under the E-books tab.

CHINA IN THE 21ST CENTURY

POLLUTION IN CHINA

MICHAEL I. CHANG
EDITOR

Nova Science Publishers, Inc.
New York

Copyright ©2011 by Nova Science Publishers, Inc.

All rights reserved. No part of this book may be reproduced, stored in a retrieval system or transmitted in any form or by any means: electronic, electrostatic, magnetic, tape, mechanical photocopying, recording or otherwise without the written permission of the Publisher.

For permission to use material from this book please contact us:
Telephone 631-231-7269; Fax 631-231-8175
Web Site: http://www.novapublishers.com

NOTICE TO THE READER

The Publisher has taken reasonable care in the preparation of this book, but makes no expressed or implied warranty of any kind and assumes no responsibility for any errors or omissions. No liability is assumed for incidental or consequential damages in connection with or arising out of information contained in this book. The Publisher shall not be liable for any special, consequential, or exemplary damages resulting, in whole or in part, from the readers' use of, or reliance upon, this material. Any parts of this book based on government reports are so indicated and copyright is claimed for those parts to the extent applicable to compilations of such works.

Independent verification should be sought for any data, advice or recommendations contained in this book. In addition, no responsibility is assumed by the publisher for any injury and/or damage to persons or property arising from any methods, products, instructions, ideas or otherwise contained in this publication.

This publication is designed to provide accurate and authoritative information with regard to the subject matter covered herein. It is sold with the clear understanding that the Publisher is not engaged in rendering legal or any other professional services. If legal or any other expert assistance is required, the services of a competent person should be sought. FROM A DECLARATION OF PARTICIPANTS JOINTLY ADOPTED BY A COMMITTEE OF THE AMERICAN BAR ASSOCIATION AND A COMMITTEE OF PUBLISHERS.

Additional color graphics may be available in the e-book version of this book.

LIBRARY OF CONGRESS CATALOGING-IN-PUBLICATION DATA

Chang, Michael I.
Pollution in China / Michael I. Chang.
p. cm.
Includes index.
ISBN 978-1-61122-022-3 (hardcover)
1. Pollution--China. 2. Pollution--Environmental aspects--China. 3. Environmental protection--China. 4. Environmental monitoring--China. I. Title.
TD187.5.C6C53 2010
363.730951--dc22

2010037902

Published by Nova Science Publishers, Inc. † New York

CONTENTS

PREFACE

China has been very successful in developing its economy in the past 30 years. However, the severe environmental deteriorations associated with the rapid economic growth have generated serious concerns. Strong efforts and significant achievements have been observed in the work of environmental protection in China. This book explores the present environmental quality and pollution emission trends in the past as well as current protection efforts in China today.

Chapter 1 - China has been very successful in developing its economy in the past 30 years. However, the severe environmental deteriorations associated with the rapid economic growth have generated serious concerns. Even though some analyses show that the economic development in China is not necessarily unsustainable after taking into consideration of its environment and natural resources, the environmental deteriorations have caused significant damages to its economy and its socio-welfare, and undermined its economic achievements. However, strong efforts and significant achievements have been observed in the work of environmental protection in China, especially in the past 10 years. The environmental institutions have been continuously improving; the investment in environmental protection has kept increasing; and the pollution intensity has kept reducing. It is fair to say that, without those efforts and achievements, the environmental quality in China could have been much worse than that of today. It should also be fair to say that the environmental efforts and achievements in China in the past were not enough, as witnessed by the unbearable pollution levels in the air, water and land, by the worrisome ecological degradations, by the economic damages and the social conflicts that the environmental problems caused to the current generation, as well as by the potential constraints that the current environmental issues post to the economic development in the future.

Chapter 2 - This study examines the status and the factors that influence how factories in Zhuji, Zhejiang negotiate China's national energy policy of reducing energy intensity by 20 percent by 2010. Using six factories as case studies, the research paper further examines the investments, modifications and motivations for change that have emerged in direct response to the national policy which aims at pollution reduction and energy reduction. Interviews conducted in the field with different company members, help to piece together a localized view of how the policy operates. Results of the field study show that ownership type effects the level of support in education and funding to reduce energy intensity in factories. Recommendations include greater education on government incentives and the fostering of stronger business networks conducive to innovation and knowledge sharing. Additionally, appropriate market-based policy mechanisms for private industry are needed to extend current implementation of energy intensity in industry.

Chapter 3 - China signed the Stockholm Convention on Persistent Organic Pollutants (POPs) on May 23, 2001 and the Chinese government is just making big efforts to phase out POPs production and consumption, eliminate POPs emissions, and dispose obsolete POPs pesticides and relevant wastes. With the requirements of convention implementation, monitoring ability and management capacity on POPs have been enhanced largely during these 10 years. Especially for the polychlorinated pibenzo-p-dioxins and dibenzofurans (PCDD/Fs), there are now more than 10 analytical laboratories established in China with high resolution gas chromatography/high resolution mass spectrometry (HRGC/HRMS) and qualified for the quantitative detection of dioxins. More and more data are published publicly, which gives us a chance to review the current PCDD/Fs levels in the ambient air, water, soil and sediment samples around China and show the trend PCDD/Fs pollution situation in these years. Furthermore, a PCDD/Fs emission analysis on the concerned industries including municipal solid waste incineration, medical waste incineration, ferrous and non-ferrous metallurgy, e-waste treatment and open burning, shows the contribution of these industries to POPs pollution in the whole country and tells the current level of pollution control technology development in China, that may be served as an academic reference for the related government departments when making a PCDD/Fs control policy in the near future.

Chapter 4 - Transportation is a leading sector for energy consumption together with associated air pollutant and greenhouse gas (GHG) emissions, and one of the most difficult sources to control. In the worldwide scope, such issues as how to reduce the consumption of non-renewable energy resources, what kind of effective remedies can be taken to mitigate air pollution and GHG emissions from transport sector have been paid more and more attentions not only by researchers but also by policy makers. Asian developing country like China, with the expected increase in levels of motorization and further economic growth, would eventually have to target air pollution control and low carbon transport more vigorously than before in the short as well as the long term.

To grasp the status of transport-related energy and environmental problems and their future trends in China so that proper policies could be made to achieve a sustainable development, in this chapter firstly the contribution of transport to air pollution in a local scale together with global warming among all the industrial sectors is investigated. Secondly, the inventories of air pollutant emissions are evaluated from inter-city transport including four modes such as railway, road, waterway and airway, and from urban transport consisting of private car, bus, trolley, taxi and rail transit. Thirdly, the strategies and policies relevant to energy conservation and emission control are summarized. More importantly, a system dynamics model is developed for quantitative policy assessment and projection of air pollution mitigation potential up to 2030.

Chapter 5 - In order to investigate indoor air quality for China, much of surveys for various air pollutants have been conducted. This chapter is focused on the pollution of aromatic volatile organic compounds (VOCs) and carcinogenic polycyclic aromatic hydrocarbons (PAHs) in indoor air in Hangzhou in China. The surveys were conducted in indoor microenvironments (living room, bedroom, and kitchen) and outdoors, which were compared to the corresponding data obtained in Shizuoka, one of urban cities in Japan. Comparing the contributions and relationships among those pollutants, the significant differences of related emission sources were clarified between both countries; throughout the samplings, the indoor and outdoor concentrations of many of the targeted VOCs (benzene, toluene, ethylbenzene, xylenes, and trimethylbenzenes) in China were significantly higher than those in Japan. The indoor concentrations of VOCs in Japan were somewhat consistent with those outdoors, whereas those in China tended to be higher than those outdoors. The concentrations of indoor PAHs in China and Japan also showed the similar trends of the case of VOCs; the level in China was extremely higher (>10-times) than that in Japan. Finally, the lifetime cancer risks estimated from unit risks and geometric mean indoor concentrations of carcinogenic VOCs were 24 × 10–5 in China and 2.6 × 10–5 in Japan. For PAHs, Toxicity potencies of PAHs in residential air of China were much higher than that in Japan. These estimations indicate that the exposure risks for air pollutants in China could be problem of deep concern.

Chapter 6 - The rapid economic growth, urban expansion, and transportation facility have led to severe air pollution problems in central-eastern China. However, very limited studies of air pollutions in this region have been conducted. In this chapter, the authors will analyze the surface ozone pollution in central-eastern China based on the measurement results at three observational sites (Jinan, an inland city; Mt. Tai, a mountain site with the altitude of 1534m a.s.l.; Yantai, a coastal city) from April 2003 to April 2006. The main results and conclusions are: (1) ozone pollution was severe in this region, especially in summer; (2) the three typical sites have obviously different characteristics of ozone pollution with each other; (3) ozone pollution at the mountain site and coastal site were both more severe than that at the urban site, which associated with the ozone depression process of urban atmosphere; (4) surface ozone present higher levels in spring and summer than that in autumn and winter at the urban site; (5) besides temperature and solar radiation, sea-land breeze circulation is an important factor influencing the ozone level at the coastal site, and maritime wind often induce high ozone levels; (6) the diurnal variation magnitude of ozone concentrations at the mountain site was much smaller than the urban site due to the lower local pollutant emissions. The study in his chapter can contribute to a better understanding of the ozone pollution in the vast central-eastern China caused by the anthropogenic activity.

Chapter 7 - China is facing the challenge of feeding its large and increasing population from a limited and decreasing area of cultivated land while achieving a clean and safe environment. After the onset of the green revolution in the 1950s, increasing inputs of synthetic fertilisers, organic manures, pesticides, and herbicides was an efficient tool to ensure the high yield in agriculture over the world. China now is the biggest user of synthetic fertilisers in the world. However this agro-chemical based intensive agriculture contributes substantially to the emission of greenhouse gases such as CH4 and N2O (Bouwman, 2001) and the entry of pollutants (nutrients, pesticide, heavy metals) into water bodies and soils. These pollutants cause adverse effects on environmental quality and public health, for example, ozone depletion in the upper atmosphere, the eutrophication in lakes and streams, the pollution of soil and food.

Chapter 8 - Agronomic crops grown on the reclaimed metal-mined wastelands are a pathway for toxic pollutants entering the human food chain. Agricultural rehabilitation of mine spoils in China is pretty common and its effect has been largely overlooked. Extensive sampling of the edible crops and associated soils have been conducted for the three typical manganese mine wastelands (Bayi, Lipu and Pingle) in Guangxi, south China and heavy metal contamination of crops was assessed against China Food Safety Standards. Simple pollution index (Pi) assessment indicated no Zn (except tea) and Cu pollution among these crops, but heavy pollution of Pb, Cd and Cr was found. Composite pollution index (Nemerow index, PN) showed 36 crops from 41 were heavily polluted with heavy metals. Peanut, soybean, Chinese chestnut, persimmon, cassava, mandarin and sugarcane were the most severely contaminated crops. Consumption of these crops may pose a health risk for humans. Crops tended to have a higher Cd accumulation (as indicated by Biological Accumulation Factor) in edible parts, thus Cd is the most important food safety threat. In terms of China Soil Quality Standard (class II), the minesoils contained much higher Cd and Cr levels, not suitable for agricultural plantation. Simple reclamation for crop plantation on minesoils is legally untenable and must be strictly controlled by the local governments. In addition, more diverse restoration goals with lower environmental risk should be encouraged for the mine wastelands in South China.

In: Pollution in China
Editor: Michael I. Chang

ISBN: 978-1-61122-022-3
©2011 Nova Science Publishers, Inc.

Chapter 1

ENVIRONMENTAL PROTECTION IN CHINA

Hua Wang[1], Hongqiang Jiang[2] and Jinnan Wang[2]
[1]Development Research Group, World Bank
[2]Chinese Academy of Environmental Planning, Beijing, China

EXECUTIVE SUMMARY

China has been very successful in developing its economy in the past 30 years. However, the severe environmental deteriorations associated with the rapid economic growth have generated serious concerns. Even though some analyses show that the economic development in China is not necessarily unsustainable after taking into consideration of its environment and natural resources, the environmental deteriorations have caused significant damages to its economy and its socio-welfare, and undermined its economic achievements. However, strong efforts and significant achievements have been observed in the work of environmental protection in China, especially in the past 10 years. The environmental institutions have been continuously improving; the investment in environmental protection has kept increasing; and the pollution intensity has kept reducing. It is fair to say that, without those efforts and achievements, the environmental quality in China could have been much worse than that of today. It should also be fair to say that the environmental efforts and achievements in China in the past were not enough, as witnessed by the unbearable pollution levels in the air, water and land, by the worrisome ecological degradations, by the economic damages and the social conflicts that the environmental problems caused to the current generation, as well as by the potential constraints that the current environmental issues post to the economic development in the future.

Major issues associated with the current environmental protection system in China rest mostly in the enforcement. Some rather advanced concepts and policy instruments for environmental protection, such as mainstreaming environmental protection, integration of environment and development, polluters' pay principle, sustainable development, cleaner production, circular economy, information disclosure, etc., have been developed and adopted in China. A comprehensive legal system and organizational structure have been set up. However, those legal and regulatory instruments are found not well implemented, and the compliance has lagged far behind. Violations of regulations are found everywhere. A lesson that one may learn from the Chinese environmental protection experiences may be that a strong participation of the general public in the environmental agenda and/or a strong political will of the central government for environmental protection are necessary in order to have an efficient balance between economic growth and environmental protection, especially when the desire for economic growth is very strong and the political system is not well advanced.

This paper first presents the environmental quality and pollution emission trends in the past as well as the protection efforts. One can see that, while the overall environmental situation is still very serious, there are indications of improvements in certain aspects and in certain areas. This paper focuses on water environment, air quality and solid waste in the past 30 years, especially the past 10 years when the detailed data are available. During the review of the trends, extensive discussions are also made on major challenges in the environmental issues. It is observed that the energy intensity of Chinese economy has been continuously decreasing and the discharge intensities of major pollutants have reduced year after year, but they have not reached the world average level. In more than three decades, China's effort in protecting the environment has been evolving and become stronger and stronger. Currently, environmental awareness is high, governmental organizations for environmental protection have been fully established, environmental laws and regulations are close to mature, and a comprehensive environmental policy system has been established. Investment in environmental protection keeps increasing. International cooperation is much stronger.

A comprehensive evaluation of all major countermeasures to environmental problems that China has employed in the past 30 years is provided in this paper. They include principles and strategies that Chinese Communist Party and the government established, laws and regulations that were enacted by Chinese regulatory authorities, as well as policies, standards and decisions that the government authorities used to manage the environment. The characteristics of Chinese environmental policy making are found to include: 1) balance between environment and economic development; 2) balance between prevention and end-of-pipe treatment; 3) use of both conventional command & control approaches and economic instruments; 4) emphases on government actions. And, the future institutional challenges are identified to include: 1) some advanced concepts are well developed but not well implemented; 2) coordination between different departments of government is weak; 3) legal system is incomplete; and 4) policy enforcement and compliance are weak. Major policy instruments reviewed include: 1) command & control measures, including Environmental planning, environmental impact assessment and "three simultaneities," total pollution load control, pollution treatment deadline, and pollution discharge permit system; 2) economic policies, including pollution levy, user charge, pollution permit trading, environmental tax, ecological compensation, green credit and green security; and 3) voluntary measures, including environmental information disclosure, environmental management system , cleaner production audit, and eco-labeling product.

Several potentially positive experiences in the work of environmental protection in China have also been identified, which include 1) The environmental zoning policy; 2) the requirements of environmental impact assessment; 3) the pollution levy system; 4) The approach of performance rating and public disclosure, 5) the community environmental dialogue.

INTRODUCTION

China has been very successful in developing its economy in the past 30 years. However, the severe environmental deteriorations associated with the rapid economic growth have generated serious concerns. Even though some analyses show that the economic development in China is not necessarily unsustainable after taking into consideration of its environment and natural resources[1], the environmental deteriorations have caused significant damages to its economy and its socio-welfare, and undermined its economic achievements. One example is that a partial estimation of pollution damage on public health and economy can be as high as 5.78% of the total GDP for the year of 2003[2]. The environmental pollution and ecological degradation in China have also caused a lot of concerns internationally. One example is that China is one of the largest GHG emitters and China's attitude towards climate change will significantly affect the global efforts and achievements in this regard.

China did spend significant efforts in its environmental protection in the course of economic development. It is witnessed by its continuous improvements in environmental institutions, regulatory policies and enforcements as well as investments in environmental protection in the past 30 years. Some of the pollution trends have been under control, and some have started improving, even though some are still deteriorating. It is believed that without strong and successful pollution control efforts in the past, the environmental situation in China would have been much worse than today, given the sheer magnitude of the vast and rapid economic growth in the past 30 years.

There are both positive and negative lessons in Chinese experiences in environmental protection. The major objective of this paper is to have a review of Chinese experiences in environmental protection and to summarize the lessons learnt. The next section of this paper will review the environmental quality trends in the past, in order to help better understand the major environmental issues that the Chinese economy generated in the past and that China is currently facing. Section 2 reviews Chinese efforts in protecting its environment in the past 30 years – the institutional improvement, the policy reform, the enforcement enhancement, as well as the investment, all of which may have implications for African countries. Section 3 evaluates the major policy instruments that the Chinese have developed and applied to the environmental protection work in China. The design, implementation, and effectiveness of the instruments are all discussed. Section 4 summarizes the lessons that can be learnt from the Chinese experiences.

1. ENVIRONMENTAL QUALITY: HISTORICAL TRENDS AND CURRENT CHALLENGES

This section presents the environmental quality and pollution emission trends in the past as well as the current challenges that China is facing. One can see that, while the overall environmental situation is still very serious, there are indications of improvements in certain aspects and in certain areas. Without successful pollution control efforts, the environmental

---

1 Roumasset et al. (2008).

2 See World Bank (2004).

deterioration would have been much more severe given the scale and the speed of the economic growth in the past. This section focuses on water environment, air quality and solid waste in the past 30 years, especially the past 10 years when the detailed data are available.

1.1 Historical Trends

1.1.1 Water Quality

Water is a serious issue that China is facing. While water shortage is an important natural problem in China, the water environmental quality is also continuously deteriorating in the past 30 years, mostly due to its heavy pollution caused by the rapid economic growth after the adoption of economic reform and opening-up polices. But some improvements in water environmental quality have been observed.

Surface Water Quality

The Chinese water system is divided into seven major river basins: Changjiang River, Yellow River, Zhujiang River, Songari River, Huaihe River, Haihe River and Liaohe River (Figure 1). According to the Environmental Quality Bulletin in the past 10 years, in the monitored sections of the seven major river basins that are monitored by the National Environmental Monitoring Network, the proportion of water quality at Grade I~III and the proportion of water quality worse than Grade V[3] have been changed significantly during the period of 2001-2007, as shown in Table 1, which indicate an improvement in the qualities.. Of the seven major water systems, Zhujiang River and Changjiang River had the best quality; Liaohe, Huaihe, Yellow and Songari Rivers had a poor quality, and Haihe River had the worst quality.

According to Table 1, China's water quality as a whole is improving. The proportion of Grade I~III, which indicate good qualities, has been rising gradually and rose about 20 percentage from the year of 2001 to 2007; the proportion of water quality worse than Grade V is decreasing gradually and dropped about 20 percentage from the year of 2001 to 2007.

Table 1. Water Quality Change in Monitored Sections of Seven Major Rivers (Unit: %)

Year	Grades I~III	Worse than Grade V
2001	29.50	44.00
2002	29.10	40.90
2003	38.10	29.70
2004	38.10	29.70
2005	41.14	26.51
2006	46.00	26.00
2007	49.90	23.60

Source: "China Environmental Quality Bulletin" (2001-2007), Ministry of Environmental Protection.

[3] "Environmental Quality Standard for Surface Water" (GB3838-2002)

Figure 1. China's Major Rivers, Lakes and Reservoirs.

In 2007, the water quality of Changjiang, Yellow, Zhujiang, Songari, Huaihe, Haihe and Liaohe Rivers was basically the same as in 2006. Of the 407 sections of 197 rivers, 49.7% is Grade I~III, 26.5% is Grade IV~V, and 23.6% is worse than Grade V respectively. The Zhujiang and Changjiang Rivers as a whole had good water quality. Songari River has been slightly polluted. Yellow and Huaihe Rivers were moderately polluted. Liaohe and Haihe Rivers were heavily polluted.

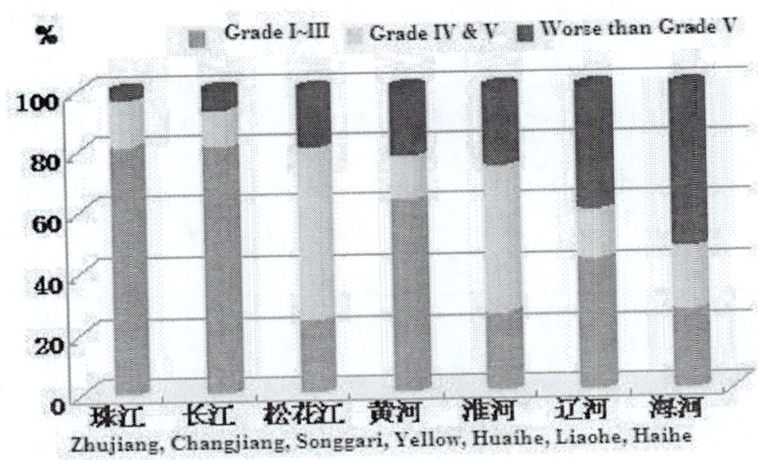

Figure 2. Water Quality Grades of Seven River Basins in 2007.

(Source: China Environmental Quality Bulletin (2007), Ministry of Environmental Protection)

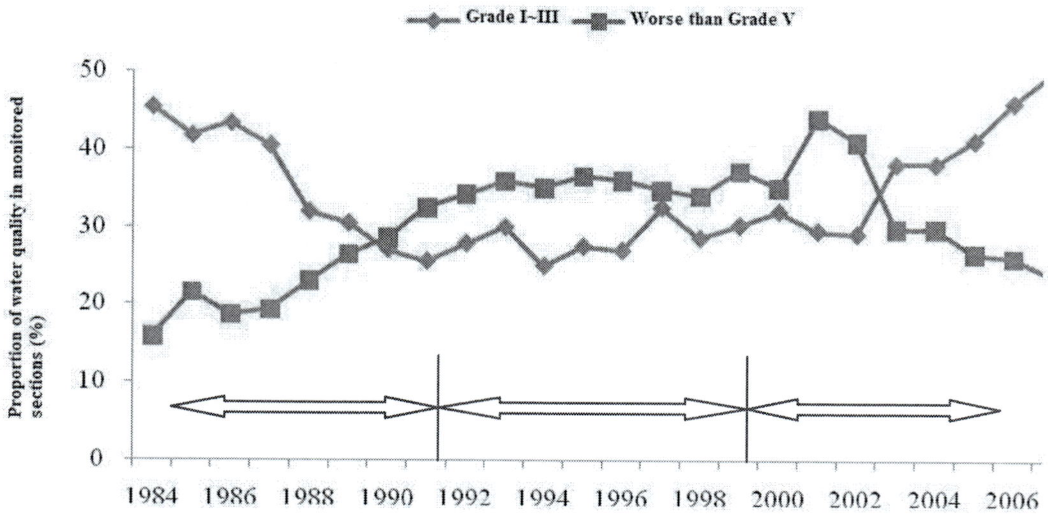

Figure 3-1. Surface Water Quality Change Trend in China in 1984~2007.

(Source: China's Environmental Condition Bulletin (1984-2007), Ministry of Environmental Protection).

Lake (Reservoir) Water Quality

According to the water quality monitoring data (presented in Table 2) of the major lakes and reservoirs, in recent 5 years, the water quality worse than Grade V in the major lakes and reservoirs accounts for over 1/3 (2003-2007) and the proportion of water quality at Grade II is also relatively low, which is only about 7%. The water quality of lakes and reservoirs in China as a whole is austere.

In 2007, for example, of the 28 nationally controlled major lakes (reservoirs), only 2 of them met Grade II water quality requirements, which accounts for 7.1%; 6 met Grade III, accounting for 21.4%; 4 met Grade IV, 14.3%; 5 met Grade V, 17.9%, and 11 was worse than Grade V, 39.3%. It was slightly better than in the year of 2006. The main pollutants were total nitrogen and total phosphorus. The water quality in the reservoirs was better than in lakes, and eutrophication in the reservoirs was only slightly.

Table 2. Water Quality of Major Lakes and Reservoirs (Unit: %)

Year	I	II	III	IV	V	Worse than Grade V
2007	0	7	21	14	18	39
2006	0	7	22	4	19	48
2005	0	7	21	11	18	43
2004	0	8	18	15	22	37
2003	0	4	21-	25	14	36

Source: "China Environmental Quality Bulletin" (2003-2007), Ministry of Environmental Protection.

Groundwater Quality

From 2000 to 2007, the number of cities with increasing groundwater pollution was usually greater than the number of cities with reducing groundwater pollution. In 2007, as analyzed from groundwater quality monitoring data of 189 cities, the groundwater quality at the main monitoring points in the monitoring zones was mainly from good to relatively poor. The quality of deep groundwater was slightly better than that of shallow ground water. The groundwater quality was better in the regions of low exploitation than in the regions of high exploitation. The groundwater quality in the countryside did not change considerably, compared with the previous years. And the regions where water quality was in a down trend are mainly found in North China, Northeast China and Northwest China. The regions where water quality was in an up trend were sparsely distributed.

1.1.2 Discharge of Water Pollutants

The major water pollutants in China's environmental statistics include COD, ammonia nitrogen and heavy metals, while the major pollution control targets in China set for water pollution control include COD, ammonia nitrogen and total phosphorus.

Discharge of Wastewater and Major Pollutants

From Figure 3-2 it can be seen that in the 11 years from 1997 to 2007, for China's discharge of wastewater and the main pollutants, the changes of COD discharge and ammonia nitrogen discharge were not the same, with nitrogen rising and COD falling. This is caused by the new pollution sources and by the technology of wastewater treatment. In general, national total wastewater discharge is basically in a trend of rising first and falling later (only discharges in 1998, 1999 and 2000 were slightly lower than in 1997). COD discharge dropped slightly and the discharge of wastewater and ammonia nitrogen rose slightly.

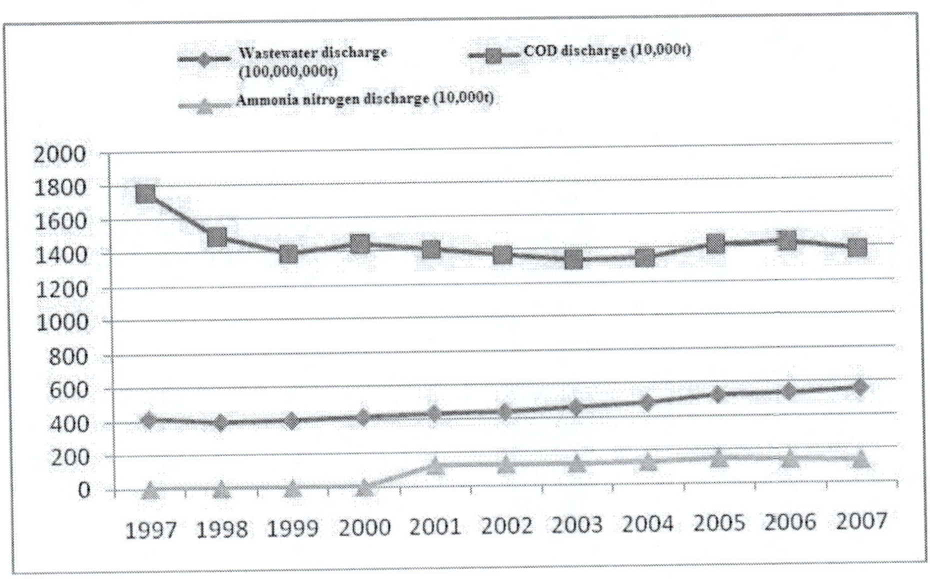

Figure 3-2. Discharge of Wastewater and Major Pollutants in China in the Past 10 Years.

(Source: China's Statistical Yearbook on Environment (1996-2007), Ministry of Environmental Protection)

In terms of wastewater discharge sources, the proportion of domestic sewage rises gradually. In 1997, the discharge of industrial wastewater was much higher than the discharge of domestic sewage. From 1998 to 2000, the discharge of industrial wastewater decreased year after year and the discharge of domestic sewage increased gradually. Since 1999, the discharge of domestic sewage (20.38 billion tons) has been higher than the discharge of industrial wastewater (19.73 billion tons). In 2007, the discharge of domestic sewage was 26% higher than the discharge of industrial wastewater.

For the discharge intensity[4], (refer to Figures 4 and 5), although the wastewater discharge and the COD discharge rise year after year, the wastewater discharge intensity and the COD discharge intensity fall year after year. The Wastewater discharge intensity (wastewater discharge per unit GDP) fell from 596.5t/10,000Yuan in 1981 to 22.6t/10,000 Yuan in 2007, and the average annual reduction rate was 12%; the COD discharge intensity fell from 22.2kg/10,000 Yuan in 1997 to 5.6kg/10,000 Yuan in 2007, and the average annual reduction rate was 13%. The industrial wastewater discharge intensity (wastewater discharge per unit of industrial value added) dropped even higher, from 1,136t/10,000 Yuan in 1981 to 23t/10,000Yuan in 2007 with an average annual reduction rate of 14%. The industrial COD discharge intensity dropped from 202.3kg/10,000 Yuan in 1997 to 47.6kg/10,000 Yuan in 2007 and the average annual reduction rate was 13.5%.

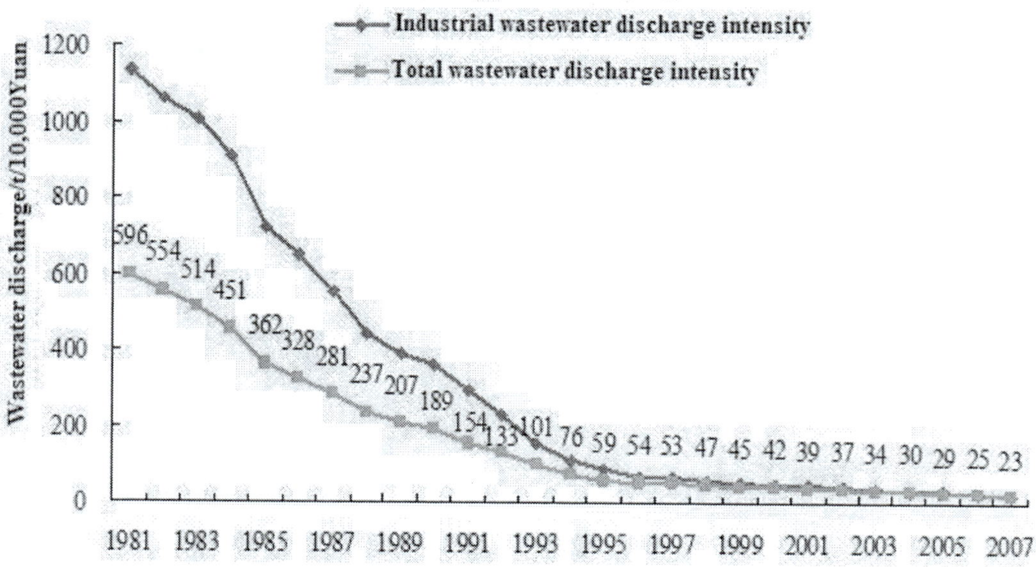

Figure 4. Total Wastewater and Industrial Wastewater Discharge Intensity in 1981-2007.

4 Wastewater discharge intensity=wastewater discharge/GDP; industrial wastewater discharge intensity=industrial wastewater discharge/industrial value added

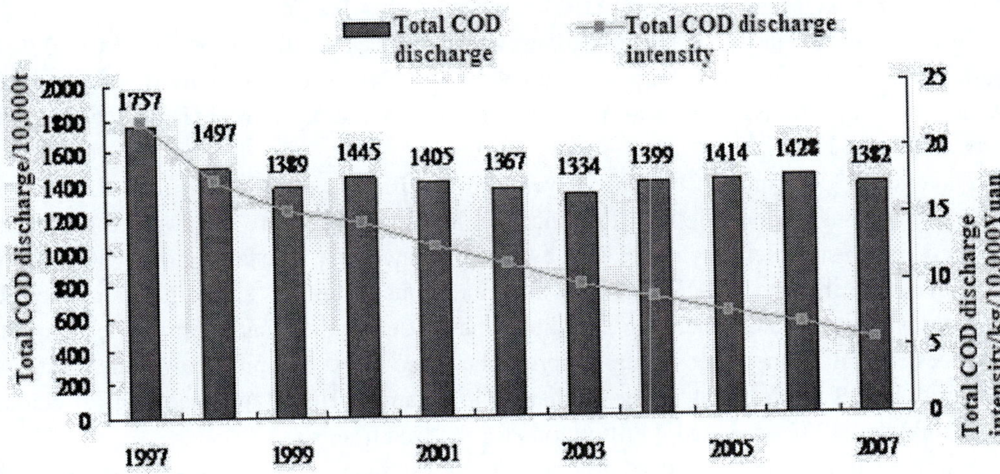

Figure 5-1. China's Total COD Discharge and Discharge Intensity in 1997-2007.

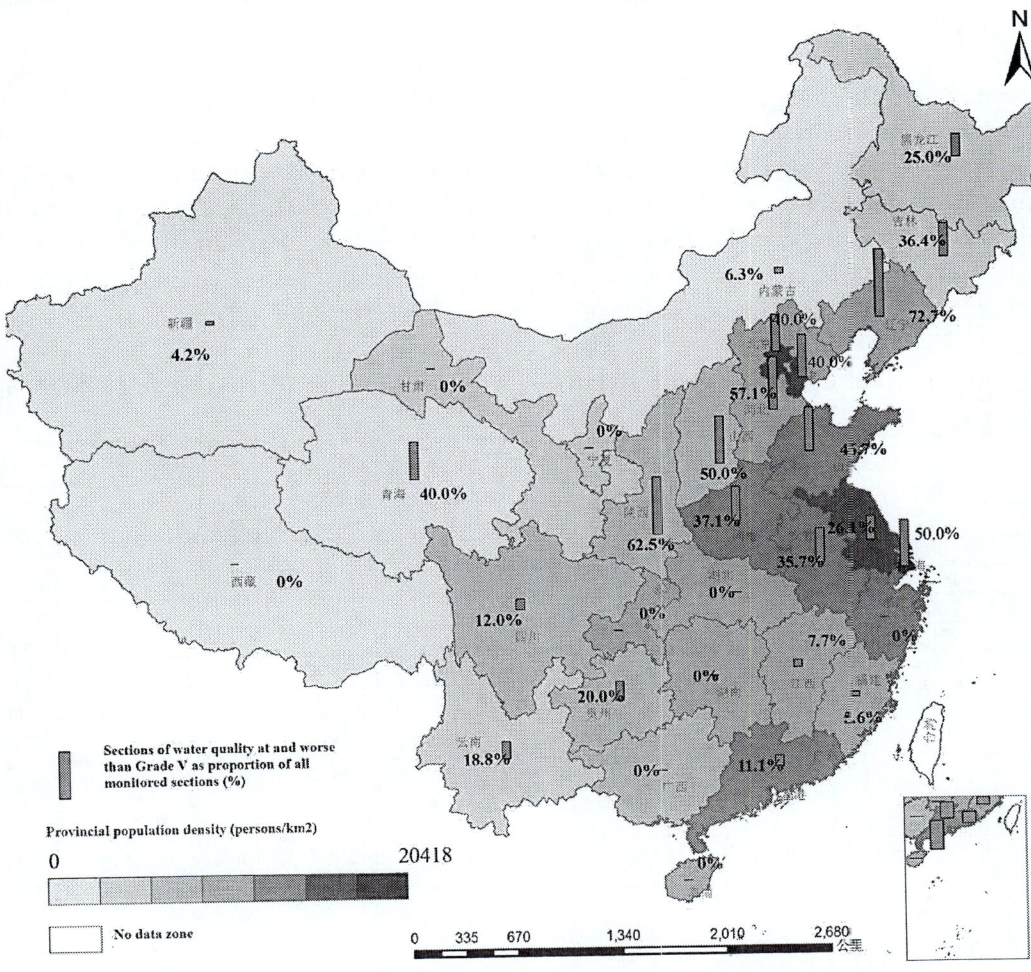

Figure 5-2. Provincial Population Density and Water Pollution Distribution in China in 2006.

Wastewater Discharge and COD Discharge in Major Regions

Environmental statistical bulletin of previous years show that from 2001 to 2006, the regions with relative high discharges of industrial wastewater and COD are mainly located in the national key drainage basins for treatment such as Liaohe River, Haihe River, Huaihe River and Taihu Lake.

Guangdong Province (Zhujiang River Basin), Jiangsu Province (Taihu Lake Basin, Huaihe River Basin) and Zhejiang Province (Taihu Lake Basin) are always in the first three positions for wastewater discharge. These three provinces are coastal regions with economy well developed, high population density and numerous industrial enterprises and therefore, the industrial wastewater discharge and the domestic sewage discharge are higher than other provinces. In addition, other booming provinces also have a rapid growth in wastewater discharge. For example, wastewater discharges in Guangxi, Hebei and Liaoning Provinces in 2003, 2004 and 2005 exceeded 2 billion tons for the first time.

The industrial wastewater discharge of the ten provinces, including Jiangsu, Guangdong, Zhejiang, Shandong, Hebei, Henan, Guangxi, Fujian, Sichuan and Hunan, has been in the top ten positions of industrial wastewater discharge in the country in the past 10 years and the accumulated contribution of wastewater discharge in these 10 provinces has been basically maintaining at 65%.

Wastewater Discharge and COD Discharge from Major Industries

Industrial wastewater discharges in China are relatively concentrated to certain sectors. From 1998 to 2001 in general, chemical industry, paper making industry and ferrous smelting industry are the major wastewater discharge sources. After 2002, the discharges from the electric power industry and the textile industry began to become the main attention while the discharge from the ferrous smelting industry has gradually decreased in percentage of wastewater discharge from key statistical enterprises (Refer to Table 7-3).

Table 3. Major Industries with Relatively High Wastewater Discharges in 1998-2006

Year	Industrial wastewater discharge (%)					
	Chemical industry	Paper making	Ferrous metal	Electric power	Textile	Total
1998	19	16	13			48.0
1999	18.8	15.6	11.9			46.3
2000	17.8	18.6	11.6			48.0
2001	17.8	16.7	10.3			44.8
2002	17.5	17.4	10.3	11.4		56.5
2003	18.1	18.4	10.3	14.6		61.4
2004	16.3	16.1	9.5	12.7		54.6
2005	15.7	17.0		11.6	8.0	52.3
2006	16.3	18.1		10.0	9.6	54.0

Source: China's Statistical Yearbook on Environment (1996-2007), Ministry of Environmental Protection

It can be seen from the industrial COD discharge figures (Figure 6) that from 1998 to 2006, although COD pollution contribution[5] in paper making and paper product industries dropped year after year, it was always in the first place; COD pollution contribution from food, tobacco and drink industries dropped approximately 10% since 2005; COD pollution contribution of chemical raw materials and product industries basically rose at small increment year after year from 1998 to 2006; COD pollution contribution of textile industry started to rise considerably in 2000 and was included as an independent industry in statistics. From 2000 to 2006, COD pollution contribution of textile industry fluctuated slightly.

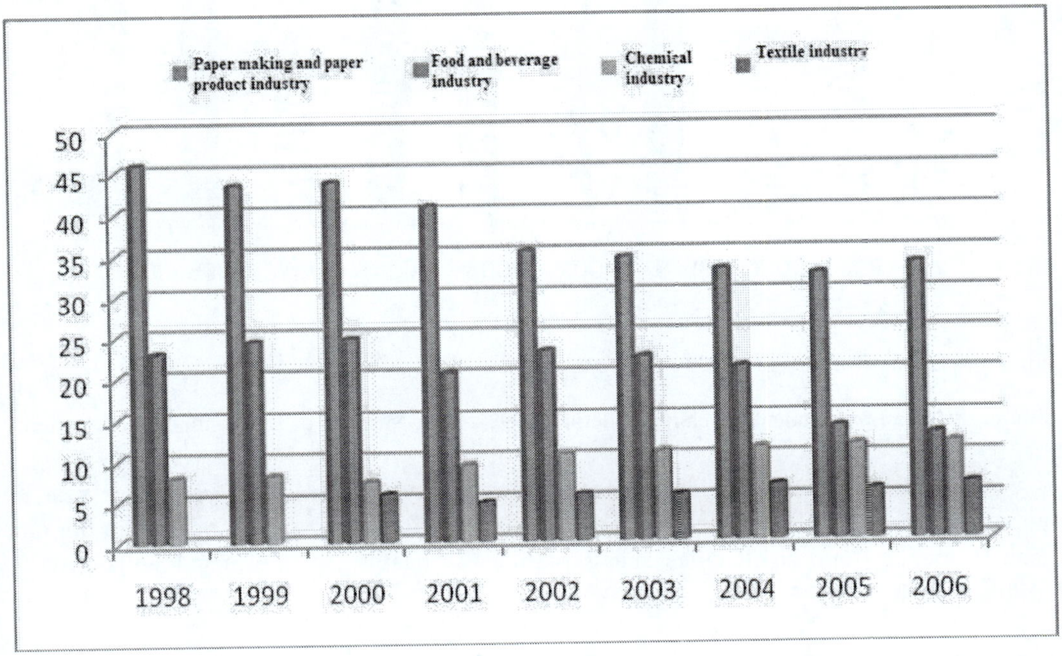

Figure 6. COD pollution contribution of major industries (%).

1.1.3 Urban Air Quality

In the past 10 years, China's urban air quality as a whole is taking a turn for the better, which is manifested by, for example, the increase of the number of cities where the air quality is at and better than Grades II and III and by the decrease of the number of cities where the air quality is worse than Grade III. (Refer to Figure 7).

In 2007, for example, of the 560 cities at and above prefecture level, 2.4% had air quality meeting national Grade I standard[6], 58.1% meeting Grade II standard, 36.1% meeting Grade III standard and 3.4% worse than Grade III. The proportion of cities worse than Grade III was down 5.7% compared with the previous year. Of the 113 major cities, 44.2% had air quality reach Grade II, 54.9% reach Grade III and 0.9% worse than Grade III. Compared with the previous year, the proportion of cities worse than Grade III was down 6.2%, much lower than 40.6% in 1999.

[5] Pollution contribution rate refers to the ratio of discharge of a pollutant from this industry to the total discharge of pollutants from statistical industries

[6] "Ambient air quality standard" (GB3095-1996)

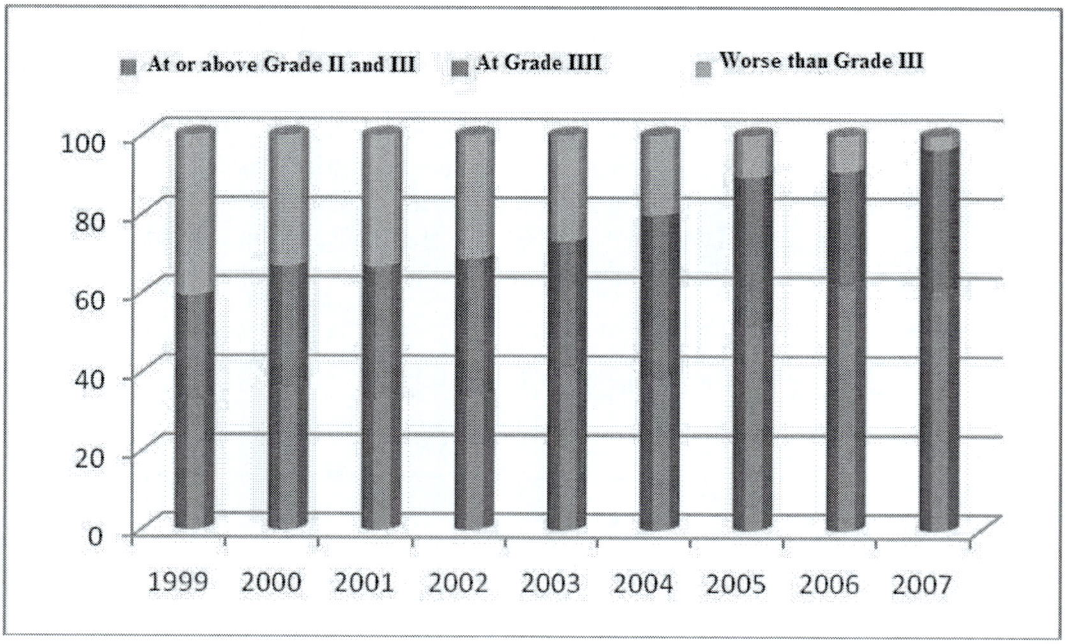

Figure 7. Trend of Urban Air Quality in China.

Source: "China Environmental Quality Bulletin" (1999-2007), Ministry of Environmental Protection.

Although the number of cities meeting air quality standard rises year after year, the population living in those cities is very low. Air pollution is apparently heavier in big cities than in medium and small cities and is particularly heaviest in the megalopolis with 1~2 million people.

1.1.4 Acid Rain

Energy consumption, especially coal consumption, grows enormously in recent years in China. Waste gas treatment projects usually have a long construction period and the emission reduction effects appear late. Consequently, the emission of sulfur dioxide and nitric oxides increases continuously in the past years, and the pollution of acid rain and sulfur dioxide is very serious in China. The heavy acid rain areas increase and the acid rain occurrence frequency rises. Pollution of fine particulates transformed from sulfur dioxide and nitric oxides is aggravated and many cities and regions show an austere situation of composite atmospheric pollution.

The acid rain monitoring results show that the precipitation acidity in the country in 1990s was steady and started to rise after the year of 2000. In 2005, the average concentration of sulfate radical and nitrate radical in precipitation was up 12% and 40% respectively. But in recent two years, the precipitation acidity has dropped slightly. Compared with 2006, the precipitation acidity in 2007 dropped slightly and the proportion of cities incurring heavy acid rain (precipitation pH<4.5) decreased by 1.3%. See Figure 8.

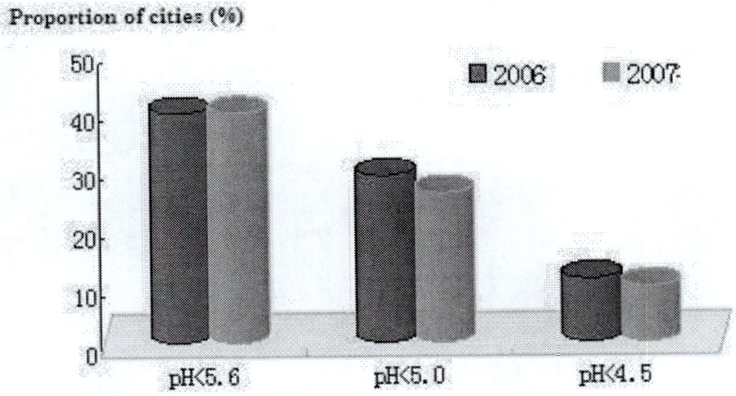

Figure 8. Proportion of Cities at Different Precipitation pHs (Average Annual Values).

The acid rain areas in China are mainly distributed in the south of the Changjiang River, including most regions of the provinces like Zhejiang, Jiangxi, Fujian, Hunan, Guzhou and Chongqing as well as part of the provinces and cities like Guangdong, Guangxi, Sichuan, Hubei, Anhui, Jiangsu and Shanghai. Acid precipitation also occurs in some of the northern regions. The heavy acid rain area increased from 4.9% of the total national land area in 2002 to 6.1% in 2005. The regions with over 25% acid rain occurrence frequency in 2006 covered 32.6% of the national land area. In 2007, the precipitation of 56.2% monitored cities was acid rain and the regions with 25% acid rain occurrence frequency covered 34.2% of the national land area; the acid rain area remained basically steady but the precipitation acidity increased. Refer to Figure 9.

1.1.5 Emission of Atmospheric Pollutants

For a long time, smoke dust and sulfur dioxide are the major air pollutants in China, due to its unique fuel-coal-based energy consumption. In 1990s, with the improvement of national economic development and people's living standard, various motor vehicles swarmed into cities in large quantities and many cities experienced the process of changing from coal burning pollution to composite pollution of coal burning pollution and motor vehicle pollution.

Emission of Major Atmospheric Pollutants

In general, emission of sulfur dioxide basically conforms to the national policy orientation. In the Ninth Five Year Plan period, China started to carry out sulfur dioxide total quantity control, and total emission of sulfur dioxide dropped from 22.66 million tons in 1997 to 19.95 million tons in 2000, down 12.0%. In the Tenth Five Year Plan period, with a rapid growth of China's energy consumption, the total emission of sulfur dioxide rebounded greatly. Total emission of sulfur dioxide reached 25.49 million tons in 2005, and reached 25.888 million tons in 2006, the highest emission in the history, up 27.8% compared with the year of 2000. According to the Eleventh Five Year Plan, China implemented the most stringent energy saving and emission reduction policies. In 2007, emission dropped for the first time in recent years to 24.68 million tons, down 3.14% from the previous year. The energy saving and emission reduction policies won their initial success.

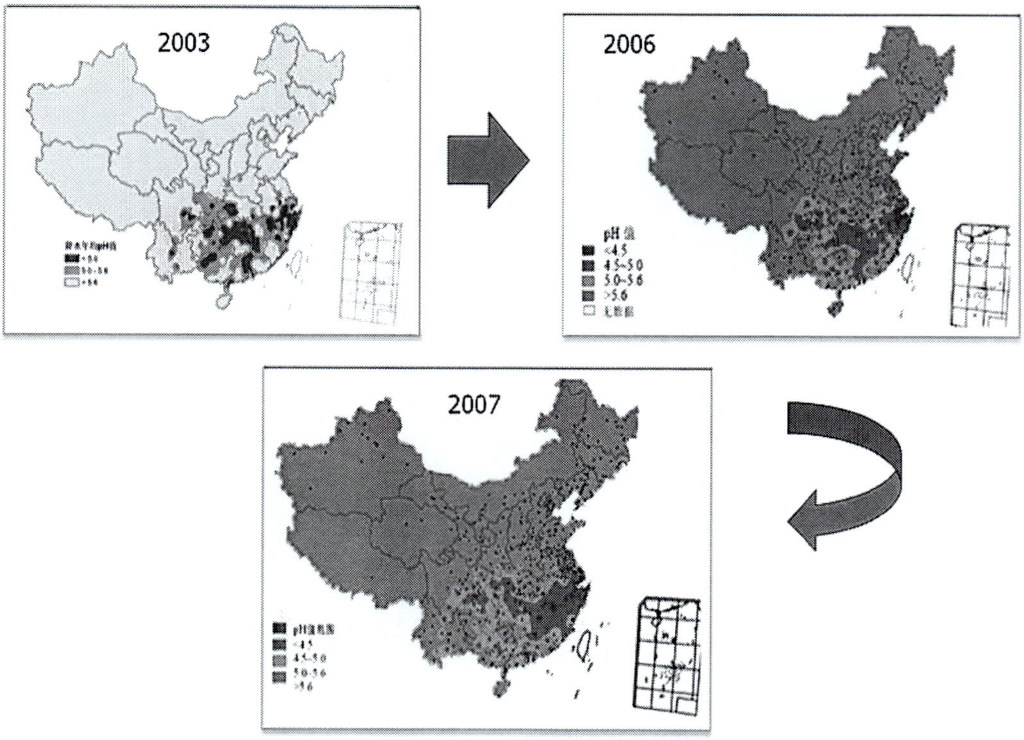

Figure 9. Acid Rain Distribution in China in Recent Years.

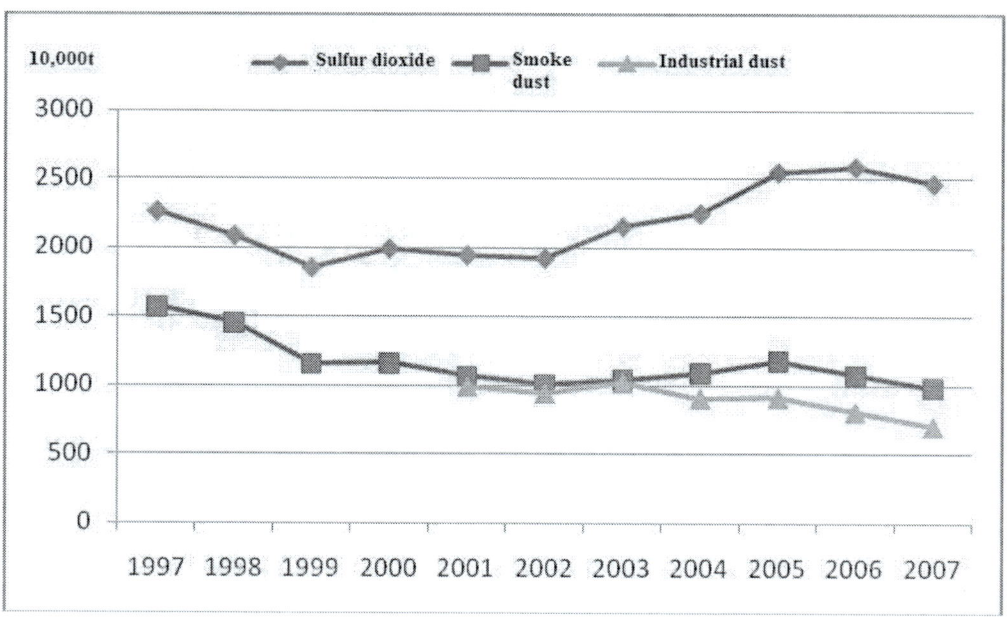

Figure 10. Emission of Major Air Pollutants in 1997-2007 in China (Unit: 10,000t).

Source: China's Statistical Yearbook on Environment (1997-2007), Ministry of Environmental Protection

In 1997-2000, the emission of sulfur dioxide and smoke dust in the country are mostly from industrial sources. (See Figure 10).

The emission intensity of sulfur dioxide (Figure 11) in China has been decreasing year after year and the average annual decreasing rate is 5.1%. The emission intensity of industrial SO2 also drops at 5.4% annually. Compared with the developed countries, however, China's SO2 emission intensity is still high and was 12 times higher than that in USA, 26 times that in UK and 78 times[7] that in Japan in 2002. In 2007, China's SO2 emission intensity was 0.068t/10,000USD, still higher than most of other countries. Given that that the coal-based energy consumption structure is difficult to change in China and that the reduction potential in emission intensity decreases gradually, it will be a heavy burden for China for quite a long time in the future to reduce SO2.

Emission of Industrial Air Pollutants

Energy production and consumption are the main sources of air pollution in China. Electric power industry, nonmetallic mineral product industry, ferrous smelting industry, chemical manufacturing industry and nonferrous smelting industry are the major emission sources of sulfur dioxide. Environmental statistics indicates that from 2000 to 2006, SO_2 pollution contribution from these industries was over 85% of the total emission from the whole industry, which is greatly higher than their economic contribution. Electric power industry is the main industry for industrial sulfur dioxide emission. Its emission grows rapidly and its proportion of industrial sulfur dioxide emission increased from 43.2% in 2000 to 58.9% in 2005 (Refer to Figure 7-12).

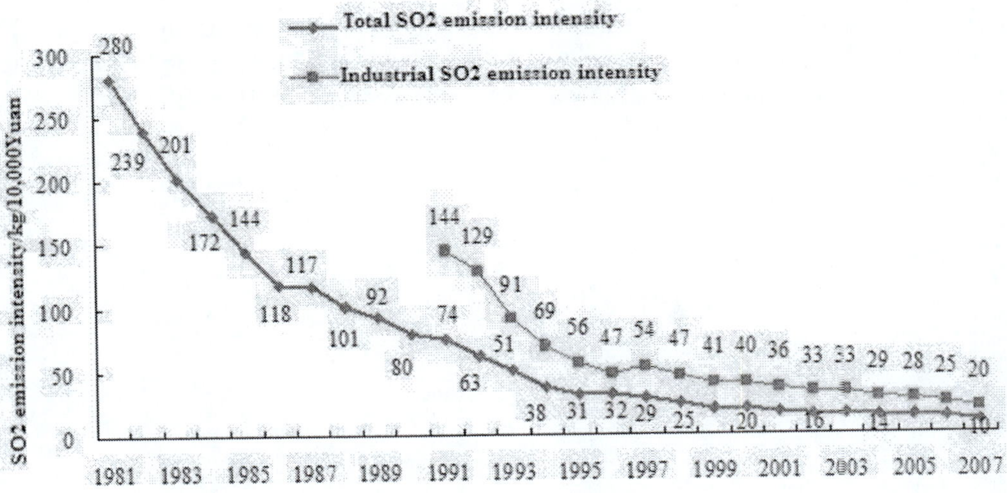

Figure 11. SO2 Emission Intensity in China.

7 "China's Statistical Yearbook on Environment 2006", National Bureau of Statistics and State Environmental Protection Administration

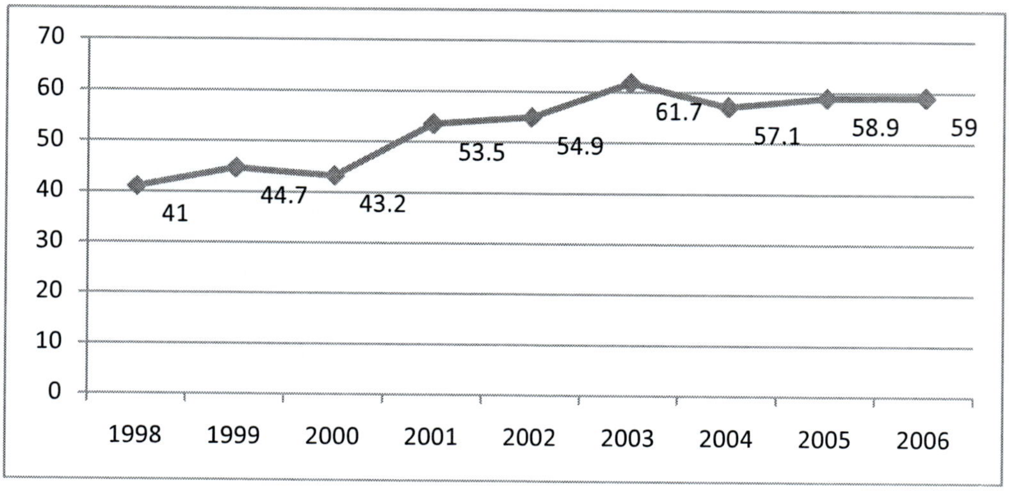

Figure 12. Contribution of SO2 Emission from Electric Power Industry, %.

Atmospheric Pollution from Traffic

Transportation development is the basis and the symbol of economic prosperity. From 1978 to 2007, the highway mileage in China increased for 303%; the civil motor vehicles increased from 1.36 million in 1978 to 43.58 million in 2007, up 31 times; the possession of private vehicles sharply increased by 28.48 million with an annual growth rate of 23% (Refer to Figure 13). The pressure of road traffic on air quality increases significantly.

With more and more motor vehicles, like in many other developed countries, the road traffic will become the primary pollution sources for NOx, CO, VOCs and fine particulates in China in the next a couple of decades. According to related research, in 2005 China's NOx emission from motor vehicles was about 6 million tons, accounting for about 1/3 of total NOx emission; VOCs emission was about 8.5 million tons; and PM10 emission from motor vehicles was about 1.3 million tons, accounting for 17% of total PM10 emission.

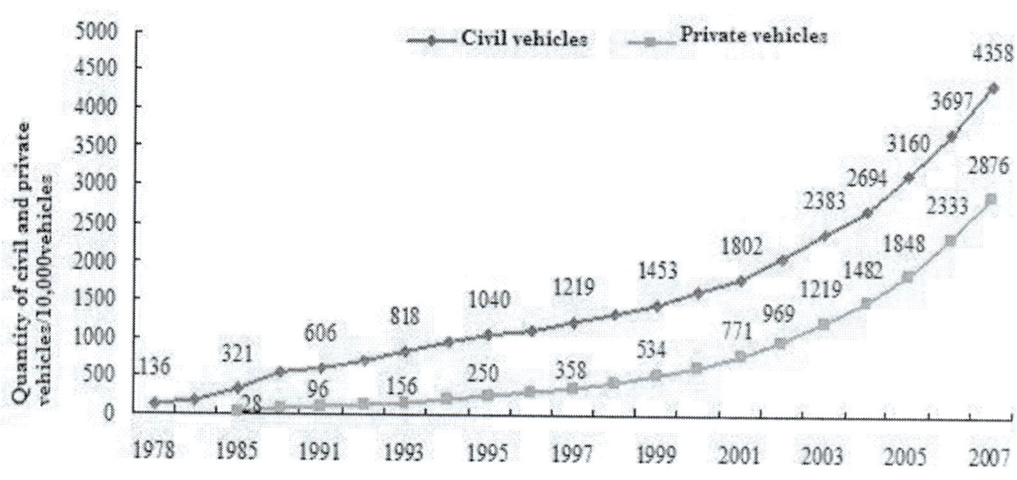

Figure 13. Possession of Motor Vehicles and Private Vehicles in China in 1978-2007.

1.1.6 Discharge of Solid Wastes

The environmental statistics shows that China's solid waste mainly has three types: industrial solid waste, hazardous waste and municipal waste.

Discharge of Industrial Solid Waste

Over the past 10 years, because of its rapid development and plus its mode of extensive consumption of resources, the production and heaping of industrial solid waste keep increasing (See Figure 7-14). In 2007 the production of industrial solid waste was 1.76 billion tons, up 16.0% from the previous year; the discharge of industrial solid waste was 11.97 million tons, down 8.1% from the previous year.

From Figure 14, the total production of industrial solid waste in 1997-2007 increased greatly and the disposal also increased. The total production in 2006 reached 2.7 times of that in 1997 and the growth rate was very high. The growth rate was the highest in 2001~2006. For the disposal method, the comprehensive utilization and disposal gradually become the main treatment methods for the solid wastes. At the same time, the storage of solid waste is gradually reduced. From Figure 14, however, even though with the increase of total quantity of solid waste, the disposal has dropped year after year since 2000 after a short period of growth, and the disposal in 2007 was less than in 1997. This is closely related to the achievement of waste management goal in the Tenth Five Year Plan for Environmental Protection.

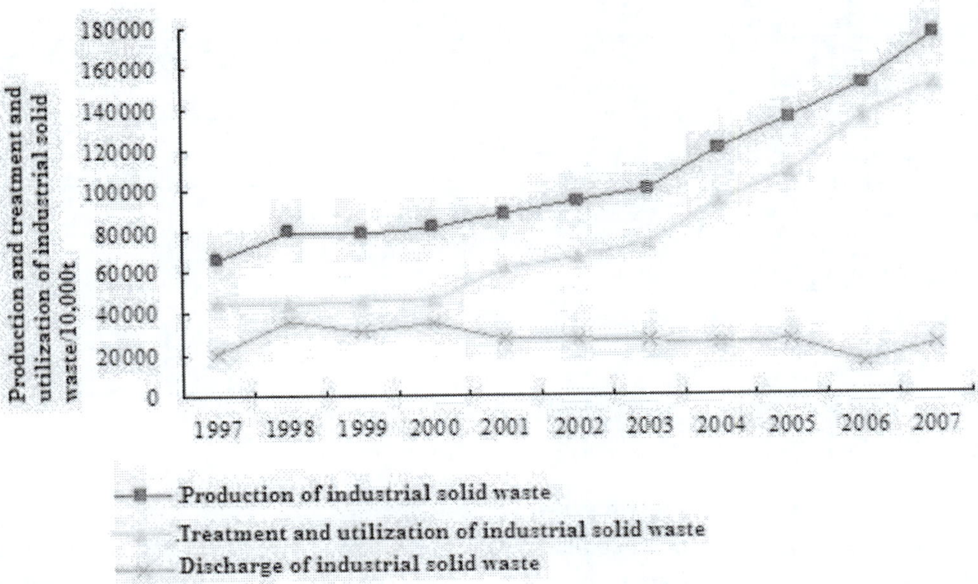

Figure 14. Production and Disposal of Industrial Solid Waste in the Past 10 Years (10,000t).

In terms of regional distribution, the discharge of industrial solid waste is mainly distributed in the high resource consuming regions. In 2006, the discharge of industrial solid waste exceeded 1 million tons in four regions: Shanxi, Guizhou, Xinjiang and Chongqing in turn. The discharge of industrial solid waste in these four regions is 63.1% of the national total discharge of industrial solid waste. In terms of industries, the discharge of industrial solid waste exceeded 1 million tons in 2006 in four industries: coal mining industry, nonferrous mining and dressing industry, ferrous smelting industry, and ferrous mining and dressing industry. The discharge of industrial solid waste from these four industries was 69.9% of the total discharge from industries.

Discharge of Hazardous Waste

It is known from Figure 15 that in the past 10 years, the production of hazardous waste fluctuated slightly but remained stable as a whole. The comprehensive utilization of hazardous waste started to increase gradually since 2004, the treatment increased slightly and the discharge decreased considerably. In 2007, for example, the production of hazardous waste in China was 10.79 million tons and was basically the same as in 1997, but the discharge decreased sharply and was 7,400 tons, only 1.64% of that in 1997. For the industries producing hazardous waste, industrial hazardous wastes in China are mainly produced in mining industry, chemical raw material and chemicals manufacturing industry, ferrous metal smelting and extrusion industry, nonferrous smelting and extrusion industry. In addition to the own features of the industries, backward mode of production, insufficient treatment equipment and lack of hazardous waste management ability are the main causes for high production of hazardous wastes from the above industries.

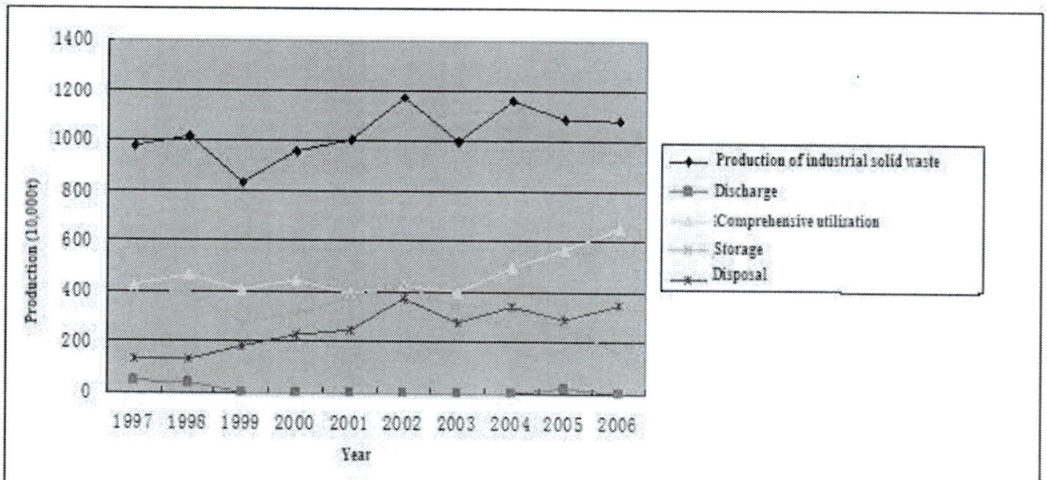

Figure 15. Hazardous Wastes in China in the Past 10 Years.

Discharge of Municipal Domestic Wastes

Municipal domestic waste collection in China is in an upward trend and this is closely related to the acceleration of China's urbanization progress, the increase of urban population and the improvement of the people's living standard. With the increase of refuse collection, sound disposal increases year after year. But the sound disposal rate is still low and the municipal refuse landfill still has a gap from the goal set in the Tenth Five Year Plan.

Table 4. Municipal Domestic Waste Discharge and Disposal in China in the Past 10 Years

Year	Domestic waste collection (10,000t)	Sound disposal (10,000t)	Sound disposal rate, %
2006	14841	7873	53.0
2005	15577	8051	51.7
2004	15509	8089	52.2
2003	14857	7545	50.8
2002	13650	7404	54.2
2001	13470	7840	58.2
2000	11820	7255	61.4
1999	11420	7232	63.3
1998	11300	6783	60.0
1997	10980	6292	57.3

Source: "China Statistical Yearbook on Environment" (1996-2007), Ministry of Environmental Protection

1.2. Major Challenges

1.1.1 New Challenges

Conventional Point-source Pollution plus Non-point-source Pollution

In the past, China's pollution was mainly manifested by its industrial point-source pollution. While the industrial pollution has not yet been brought under control, the non-point-source pollution has also become a serious problem. Especially in East China, due to the excessive and irrational use of chemical fertilizers and pesticides, the rapid development of centralized livestock and poultry breeding industry as well as the increase of urban and rural domestic sewage discharge, the non-point-source pollution has increased dramatically. In the urban areas, with the improvement of quality of life, the retention of motor vehicles grows rapidly, and the vehicle emission has become the non-point-source pollution that is very difficult to deal with in China.

Domestic Pollution

Due to the improvement of living standard and change of consumption pattern, urban and rural domestic wastes and domestic sewage grow quickly and have become a difficulty in China's environmental treatment. In 2007, the urban sewage treatment rate in China was only 62.87% and the reutilization rate of urban sewage was 7%. And with use increase of numerous chemical products, the waste composition also has changed Chemicals and nutrient contents in water increased, which increases the difficulty of wastewater treatment and makes eutrophication of lakes, reservoirs and offshore water areas stay high.

New Pollution Problem

With the increase of pollution treatment, some traditional pollutants are under control to some extent. But new pollution issues keep emerging. For example, with the increasing attention to global environment and the continuous appearance of climate change effect, the emission of greenhouse gases has been gradually incorporated into the country's environmental agenda.

Pollution Migration

High pollution industries tend to transfer to less developed areas. Due to the unbalanced development between the regions, some heavy pollution industries in the developed regions are transferring to the less developed regions. The ecological and environmental damage in the economically underdeveloped regions tends to aggravate, and this problem is becoming more and more serious. In the relatively developed areas, with industrial restructuring and upgrading as well as more stringent environmental polices and measures, the polluting enterprises with backward production capacities gradually lose their competitive capability. However, in the economically backward regions, the demand and the desire for development are stilly intense and those high polluting enterprises currently located in the developed regions can find chances in the less developed areas.

Ecological Problem

After the catastrophic flood in the Changjiang River in 1998 and the frequent sandstorms several years ago, China's ecological problem and especially the ecological deterioration problem in the western regions are found to be more and more serious. The land deterioration, biodiversity reduction, water resource scarcity and forest quality reduction are in an upward trend, and their severity, coverage of influence and restoration difficulty make them in a more and more serious position on China's environmental agenda. The ecological problem can not only lead to deterioration of natural conditions for the regional development, but also can cause wide-range ecological unbalance, exacerbate poverty, disaster risk and ecological crisis, make it difficult for economy to grow continuously, and trigger social instability. Although the western development program lists ecological and environmental construction a priority area, it still remains as a high risk.

1.1.2. Observation and Discussion

Water and Air Pollution

Water pollution and air pollution are still of the top priorities in China. Among the numerous environmental problems that China is facing, water pollution is the most serious and difficult one. China started to establish the water quality monitoring system in 1984. In 24 years, China's surface water environmental quality as a whole is in a deterioration trend. The spatial distribution of water resources, production and population poses a high pressure to water environment in Northern China. In 2006, for example, in terms of the pollution density and the proportion of sections of water quality at or worse than Grade V (Figure 7-17), water quality is significantly deteriorated in the north regions, especially in the Haihe River Basin, the middle and lower reaches of the Yellow River and the Northwest inland regions.

New types of atmospheric pollutants keep increasing and the regional air quality is increasingly deteriorated. From the late 1990s, with the adjustment of urban industrial structure, the increase of urban centralized heating and gas supply capacity and the large scale of installation and operation of dust and sulfur removing facilities, urban air pollution in China has been considerably improved compared with the early 1990s. But as the pollution sources moved from cities to suburbs and countryside and the pollution type changes from coal burning pollution to composite pollution of coal burning and motor vehicle emission in the urban areas, China's air pollution is spreading from local urban areas to regional areas. The effects of fine particles in the air and new types of pollutants like ozone emerge quickly. The acid rain pollution is aggravated and is spreading. The treatment of toxic and hazardous waste gases is not timely. The north regions are also affected by sandstorms. It can be seen from satellite photographs that in the east coastal regions in China, composite pollution regions such as Changjiang Delta, Zhujiang Delta, Beijing-Tianjin-Hebei, Huanghuai Plain and Central and South Liaoning Province have formed, with brown clouds emerging, which are more harmful and difficult to handle.

Comprehensive Utilization and Management of Solid Waste

Comprehensive utilization level of industrial waste residues in China is still relatively low. Presently, the total recovery rate of mineral resources in China is only 35%, 20% lower than world average. Comprehensive development of complex ore and associated ore is only 1/3. If the total recovery rate of mineral resources in China is increased by 10%, the development and extraction of mineral resources can be reduced for at least 1/4. At present, numerous utilizable solid waste resources are wasted as the "three wastes" and can not be sufficiently utilized. With regard to hazardous waste management, China's management system is unsound; its ability is weak and no unified supervision and management system has been established. Many places do not have enough supervision and management instruments or strictly enforcement of laws, which leads to a complicated flow of hazardous waste.

Ecological Deterioration

In general, the ecological quality in those big river systems is decreasing day by day and the ecological functions in up stream water conservation areas are seriously deteriorated. The vegetation in those important northern windbreak sand fixation zones is severely damaged, and the sandstorms occur frequently. The ecological system in the river flood regulation and storage zones is deteriorated. Wetlands are reduced and the functions deteriorated. The forest quality is low and the ecological regulation function has dropped. Biodiversity decreases and the resource utilization activities severely damage the ecological environment.

Water ecology is unbalanced. China's per capita water possession is only ¼ of the world average and the development and utilization of some rivers have exceeded the international warning line. Water utilization in the Yellow, Huaihe and Liaohe Rivers has exceeded 60% and in Haihe 90%. The groundwater level in some regions has dropped and different sizes of groundwater depressions are formed.

Land is seriously deteriorated. The area of soil erosion in the country is 3.56 million km2 and the area of desertification is 1.74 million km2. Although six major forestry projects have been implemented and the land desertification trend is slowed down, the desertified land is still extensively distributed in the northern arid and semiarid regions. Water erosion, wind erosion, soil salinization and soil pollution coexist.

Ecological quality in rural areas has dropped significantly. In recent years, with rapid development of rural economy, discharge of rural domestic wastewater, refuse and waste from agricultural production and livestock and poultry raising increase year after year. "Dirtiness, disorderliness and badness" is very common in rural areas, and the environmental condition of rural regions is more and more deteriorated, which directly threatens the living environment and health condition of big rural population.

Biodiversity is reduced sharply. The quality and the management level of the existing nature reserve constructions are low. The speed of species' endangering and extinction is increasing. Biological heritage resources are severely eroded.

Economic Growth

What China took in the past 30 years is a typical path of "high economic growth, high resource consumption and high environmental cost." According to the estimation made by some Chinese research institutes, in mid-1990s, two thirds of China's economic growth came from its ecological environmental overdraft. In the twentieth century, Chinese research institutions estimated that the loss from environmental pollution and ecological damage in the early twentieth century was about 7-8% of national GDP annually.

Public Health

The threat of environmental pollution against public health is mainly from water pollution, atmospheric pollution and soil pollution etc. There are a number of reports on the rise of disease incidence rate and mortality caused by pollution in China. For example, 37 types of POPs have been detected from surface water of the Taihu Lake and the concentrations of pollutants in fish bodies are much higher than that in water. Statistics shows that the mortality related to water pollution is high in many areas of China; water pollution caused hundreds of deaths in some villages in Henan, Guangdong and Shandong; and over 20 villages on the bank of Huaihe River are called as "Cancer Villages."

A Big Gap from the World Class

The energy intensity of Chinese economy is significantly reduced but there is still a gap from the world average level. The coal consumption intensity in 2007 was only about 1/13 of that in 1978, and the energy consumption intensity has reduced for 73%. However, compared with other countries in the same period, China's energy consumption intensity is still very high. Even in recent years when energy conservation is encouraged and structural adjustment is operated, China's energy consumption intensity is still 3 times the world average level. In the Eleventh Five Year Plan period, China has proposed the goal of 20% reduction in energy consumption intensity; this may drive China's energy consumption intensity to further reduce and close to the world average level.

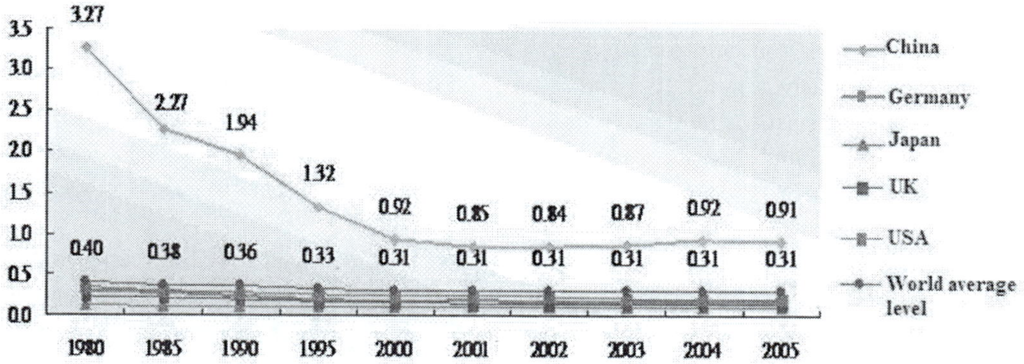

Figure 16. Comparison of Energy Consumption Intensity between China and Other Countries.

On the other hand, the discharge intensities of major pollutants have reduced year after year but have not reached the world average level. In the past 30 years, the discharge of major pollutants has been in a continuous upward trend but the discharge intensity of various pollutants dropped sharply. The wastewater discharge intensity has reduced from 236.8t/10,000Yuan in 1988 to 22.6t/10,000Yuan in 2007, down more than 10 times; the COD discharge intensity has reduced from 0.157t/10,000Yuan in 1986 to 0.006t/10,000Yuan in 2007, only 1/28 of that in 1986; the emission intensities of sulfur dioxide and smoke dust have also reduced by 96% and 99% respectively. Compared with developed countries, however, China still has a big gap. China's discharge intensity of sulfur dioxide in 2007 was 0.01t/10,000Yuan while the discharge intensity of sulfur dioxide in Japan, Germany and UK in 2002 was 0.002, 0.003 and 0.006t/10,000Yuan[8] respectively.

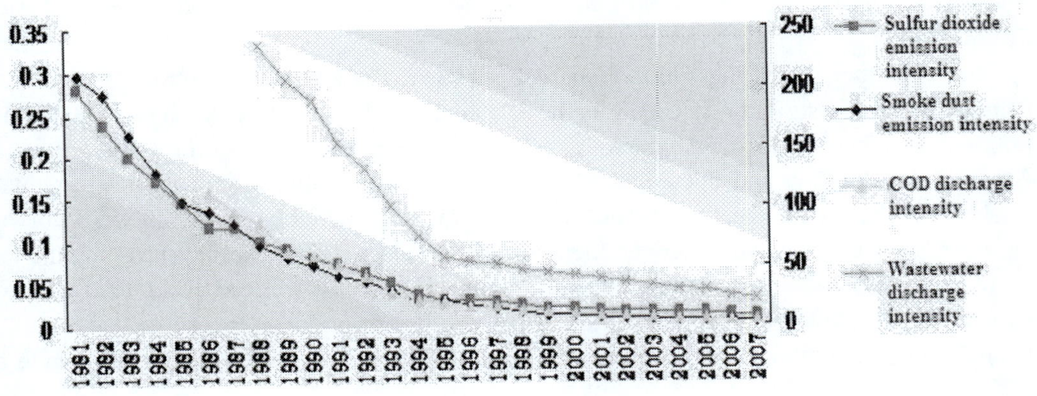

Figure 17. Discharge Intensity of Major Pollutants in China .

2. PROTECTION EFFORTS: EVOLUTION AND MAJOR ACHIEVEMENTS

2.1 Evolution

In more than three decades, China's effort in protecting the environment has been evolving and become stronger and stronger. Roughly the environmental protection work can be divided into the following four stages:

Stage 1: Beginning Stage (1972-1978)

China's environmental protection course started in 1972 when the government sent its delegation to attend, as an observer, the first conference on human environment organized by the United Nations in Stockholm, the capital of Sweden. It was through this conference that the top policy makers recognized that China also had serious environmental problems. Then, the State Council held the first National Conference on Environmental Protection which enacted a 32-word strategy for environmental protection: "comprehensive planning, rational layout, comprehensive utilization, changing disadvantages to advantages, relying on people,

[8] "China's Statistical Yearbook on Environment 2006", National Bureau of Statistics and State Environmental Protection Administration

everybody participation, protecting environment and benefiting the people," passed China's first environmental protection policy document "Several Decisions on Protecting and Improving Environment," established the State Council's Leading Group for Environmental Protection. After the conference, all provinces, cities, autonomous regions, and State Council's various departments successively established environmental management units and scientific research and monitoring institutions. During 1973-1978, China carried out pollution investigation in key area and started to formulate the national environmental protection plans. In this beginning stage, pollution prevention and control work was focused on the treatment of "three wastes" and waste utilization. Pollution management measures, such as the "three simultaneities", pollution levy, and environmental impact assessment, were established and implemented.

Stage 2: Comprehensive Development of the Environmental Management System (1978-1992)

In December 1978, the Third Plenary Session of the Eleventh Central Committee of China's Communist Party (CCP) was held, which made a historical change in China's political and economic development path. Economic polices of reform from a planned economy to a market and opening up to foreign investment and trade have been gradually established and implemented. The environmental protection work moved forward accordingly. The party and the government gradually recognized the arduousness and the long term nature of the work on environmental protection. They were actively learning from experiences in the developed countries and constantly exploring on measures that would fit China's own conditions. The major institutional achievements during this stage include the following aspects:

First, the understanding of environmental problems and pollution control strategies were significantly improved. In 1978, the Central Committee of CCP approved the policy document "Key Points of Environmental Protection Work Report" submitted by the State Council's Leading Group for Environmental Protection and pointed out that "Eliminating pollution and protecting environment is an important integral part of constructing socialism and realizing the four modernizations......We must not make detours by development first and clean-up later. We shall resolve the environmental pollution problems at the same time of construction." This was the first instruction on environmental protection by the Party Central Committee in the history of the Chinese Communist Party. It drew attention from the Party organizations at all levels and strongly pushed forward the development of China's environmental protection work. In 1983, the State Council held the second National Conference on Environmental Protection which clearly proposed that environmental protection should be a fundamental national policy and confirmed the important position of environmental protection in national development agenda.

Second, the use of legal approaches to environmental protection started. In 1978, the First Session of the Fifth National People's Congress made a provision for environmental protection in the "Constitution" for the first time which laid a solid foundation for construction of environmental laws. The "Environmental Protection Law (Trial)" was passed in 1979 and the "Environmental Protection Law" was formally promulgated in 1989.

Third, the environmental management system was preliminarily established. In the Second National Conference on Environmental Protection held in 1983, the strategies such as the "Three Synchronous Steps" (planning, implementation and development at the same time) and the "Three Consistence " (consistent between economic, social and environmental benefits) were proposed, and three environmental principles, i.e. "prevention first, coordination between prevention and treatment, and comprehensive treatment", "polluters to clean-up" and "strengthening environmental management", were determined. In 1989, the State Council held the third National Conference on Environmental Protection, which proposed five new environmental management policies and measures, which include environmental protection target-responsibility system, quantitative evaluation system for comprehensive treatment of urban environment, pollutant discharge permit system, centralized pollution treatment and control, and pollution treatment deadlines. The environmental management work became more scientific and better institutionalized.

Fourthly, the environmental management organizations were enhanced. In 1982, the Ministry of Urban and Rural Construction and Environmental Protection was established which hosted a Bureau of Environmental Protection. In 1984, the State Council established the State Council's Environmental Protection Committee that led, organized and coordinated the national environmental protection work. In 1988, National Environmental Protection Agency was established as an independent agency at the vice-ministry level directly under the State Council. The national environmental protection institution kept reinforcing, and environmental protection was placed in a more and more important position.

Stage 3: Enhancement of Environmental Protection (1992-2008)

United Nations Conference on Environment and Development was held in Rio de Janeiro in 1992, and the adoption of sustainable development strategies have become a consensus. The Fourteenth and Fifteenth Central Committee of CCP proposed to implement the sustainable development strategy and called for two fundamental shifts: shift from a planned economy to a socialist market economy; shift from an extensive economic growth model to an intensive economic growth model. With the reform of the socio-economic system, environmental protection work has also been transformed gradually in this period, in the following ways:

1) Industrial pollution control measures were modified in three aspects: from end-of-pipe abatement to whole production process control; from pollution concentration control to a joint control of total load and concentration; from scattered treatments to centralized abatement. In 1993, the Second National Working Conference on Industrial Pollution Prevention and Control was held in 1993. It summarized the experiences and lessons in the past and proposed to carry out cleaner production measures and to change the industrial pollution control strategies. Since then, industrial pollution control in China entered a new era.

2) The environmental protection work became more scientific and better met the principles of sustainable development. In 1992, China took the lead in proposing the "Ten Major Countermeasures for Environment and Development", which for the first time proposed to change the traditional development model and take the sustainable development path. China also formulated such programmatic documents as "China Agenda 21" and "China Action Plan for Environmental Protection." Sustainable development strategy became the basic guiding ideology for China's economic and social development. In 2005, the Fifth Plenary Session of the Sixteenth Central Committee of CCP proposed to accelerate the construction of resource saving and environment-friendly society and form the resource saving growth model and the sound and civilized consumption model in the whole society. In 2005 the State Council issued a document "The Decision on Implementing the Scientific

Development Outlook and Strengthening Environmental Protection," which proposed to use the scientific principles to lead environmental protection work and coordinate socio-economic development and environmental protection. In April 2006, the State Council held the Sixth National Conference on Environmental Protection and Premier Wen Jiabao emphasized the environmental protection work. In 2007, the Seventeenth CCP National Congress integrated ecological civilization into its political report for the first time, included "building the resource saving and environment-friendly society" into Party's Constitution, and set up, for the first time, "building ecological civilization" as a strategic task and an objective for building all-round well-off society.

3) Overall environmental protection strategies were adjusted and new ones were innovated. In 1996 the State Council held the Fourth National Conference on Environmental Protection and issued the "Decision of the State Council on Several Issues Concerning Environmental Protection" and defined the objectives, tasks and measures for trans-century environmental protection work. This conference determined to adhere to the policy of paying equal attention to pollution prevention and control and to ecological protection, and initiated two action plans: "Plan for total pollutant load control" and "Plan for trans-century green construction projects". Big- scale projects for pollution control and ecologic protection started since then in major cities, regions, river basins and sea shores. In March, 2006, the Fourth Session of the Tenth National People's Congress approved the "Outline of the Eleventh Five-Year Program for National Economy and Social Development," which required accelerating the construction of the resource saving and environment- friendly society, reducing energy consumption intensity of the national economy by 20% and total discharge of major pollutants by 10% in 2010 from the values in the year of 2000, and incorporating the targets of reducing pollution emission and energy consumption into the comprehensive evaluation system for economic and social development and into the comprehensive performance evaluation of government and enterprise leaders.

4) Environmental protection institutions were further strengthened. In 2007, the 17th National Congress of CCP proposed to "increase institutional integration and explore the functionally unified super-ministry system" and defined the direction for future institutional reform. The Third Plenary Session of the 17th Central Committee of CCP proposed the idea of establishing Ministry of Environmental Protection. In March 2008, NPC decided to establish Ministry of Environmental Protection, a member of the cabinet, in order to better coordinate policies, plans and major actions related to environmental protection.

Stage 4: New Era (2008-present)

With the establishment of the Ministry of Environmental Protection and given the current status of environmental severity and social and political forces for environmental protection, a new era for environmental protection is expected to come. The current institutional status will be summarized in the following section, which lists the major institutional achievements up to 2008 in the area of environmental protection, and the improvements in the future will also be discussed in the next section.

2.2.2 Major Institutional Achievements and Current Status

Environmental Awareness is High.

China's extensive growth model brought about heavy environmental pollution and ecological damage which have caused serious impacts and threats to people's life and the environment. Gradually, the country paid greater and greater importance to the environmental protection work and the government set environmental protection as a fundamental national strategy. Environmental awareness has become high among the ordinary citizens, due to the long-lasting serious environmental problems as well as the extensive environmental education and propaganda made both by the government agencies and by the non-governmental sectors. The levels of environmental ethics and morality keep progressing.

Governmental Organizations for Environmental Protection Have Been Fully Established.

China has now well established its governmental management system for environment, with State Council's overall leadership and Ministry of Environmental Protection's overall monitoring and supervision. All ministries are responsible for environmental work within their own sectors, and local governments are responsible for environmental quality in their own regions. Currently all four levels of government - national, provincial, municipal and county level, have their respective environmental departments, and some township governments in well developed areas have also established their environmental departments. By the end of 2007, the environmental protection system nationwide had in total 11,932 governmental institutions for environment and had 176,988 environmental protection personnel at all levels. The quality of the environmental protection workers and officers keeps improving. About 5.4% of the environmental personnel hold senior technical titles, 14.6% hold medium level titles and 19.2% hold primary technical titles. Over 75% hold educational degrees of college level or above.

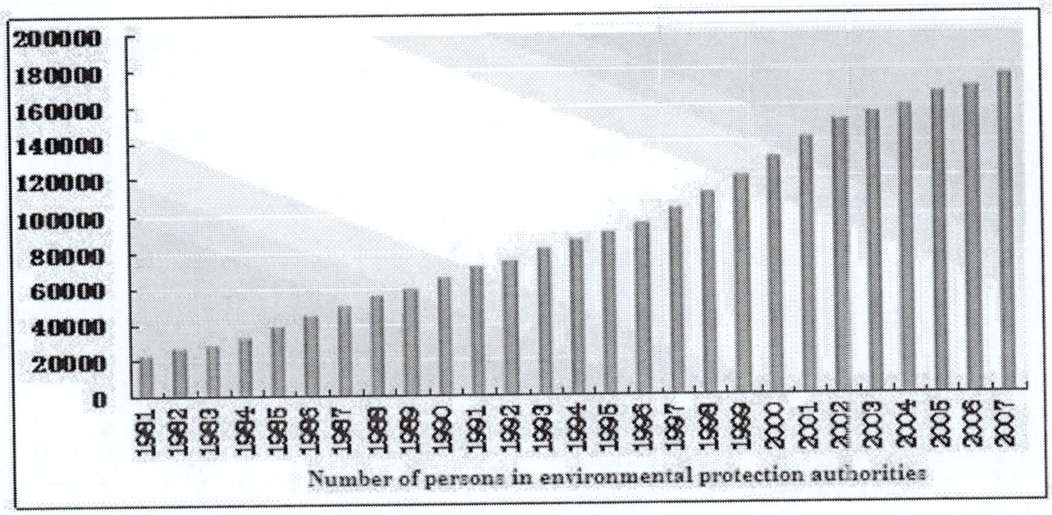

Figure 18. Number of Persons in Environmental Protection Authorities (Persons) .

Environmental Laws and Regulations are Close to Mature.

In 1989, China issued formally its "Environmental Protection Law." Since then, China has enacted 9 environmental protection laws, over 50 regulations, about 200 administrative rules for environmental protection, and more than 500 environmental standards, and has approved and signed 51 international conventions or agreements on environmental and resource protection. There are also more than 1,600 local laws and regulations.

A Comprehensive Environmental Policy System has been Established.

In the past, due to the prolonged adoption of planned economic system, China was controlling its environmental problems mainly by the administrative approach. Since the economic reform and opening up, with the steady advancement of the market economic system, China's environmental policies also started their corresponding transformation, from a simple administrative order to a comprehensive use of multiple environmental management policy instruments including command and control, economic and technical instruments. In the development course, the governmental direct control is continuously reducing, the role of government becomes more directive, and the enterprises gradually make active decisions.

Investment in Environmental Protection Keeps Increasing.

China's investment in environmental protection increased year after year, and the last decade is a period in which the increase of investment in environmental protection was the fastest. In 2007, environmental investment in China, mostly for pollution prevention and control, reached 338.4 billion Yuan, or 1.37% of its GDP. Table 7-19 shows that that China' investment in environmental protection is continuously rising.

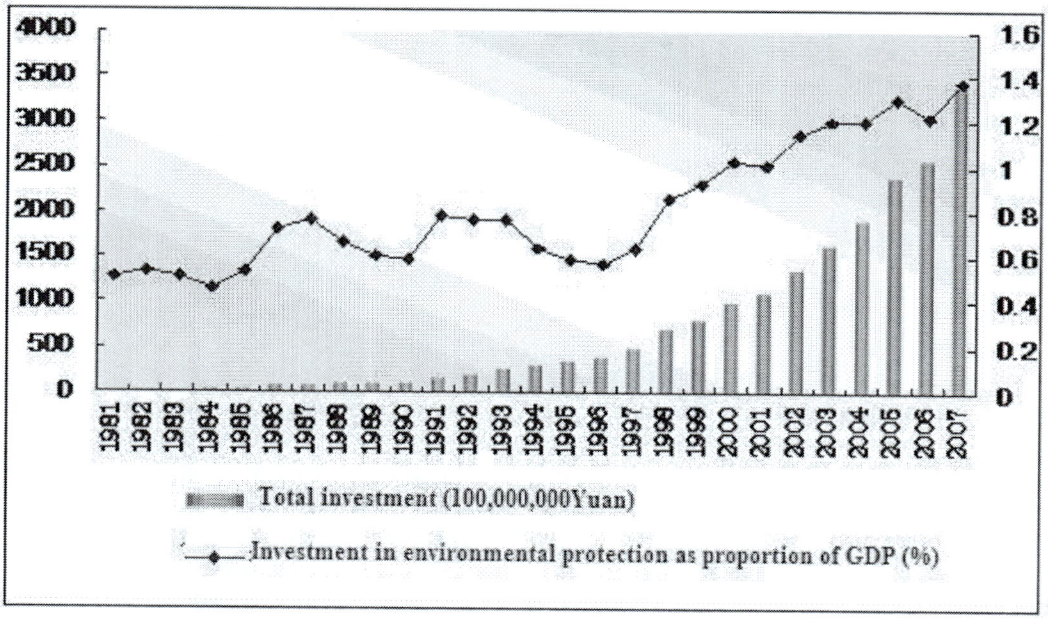

Figure 19. China's Environmental Investment and its Proportion in GDP.

International Cooperation is Much Stronger.

China has actively participated in the environmental activities initiated by such international organizations as the United Nations, and has been active in international environmental conventions. Up to now, China has concluded or signed more than 50 international environmental conventions. It has strengthened its cooperation with surrounding countries and related regions and actively participated in regional cooperation mechanism constructions. It is also active in its bilateral cooperation in the environmental protection area. Up to present, China has signed bilateral environmental protection cooperation agreements or memoranda of understanding with 42 countries and signed nuclear safety cooperation bilateral agreements or memoranda of understanding with 11 countries.

3. Policy Evaluation

3.1 Overall Evaluation

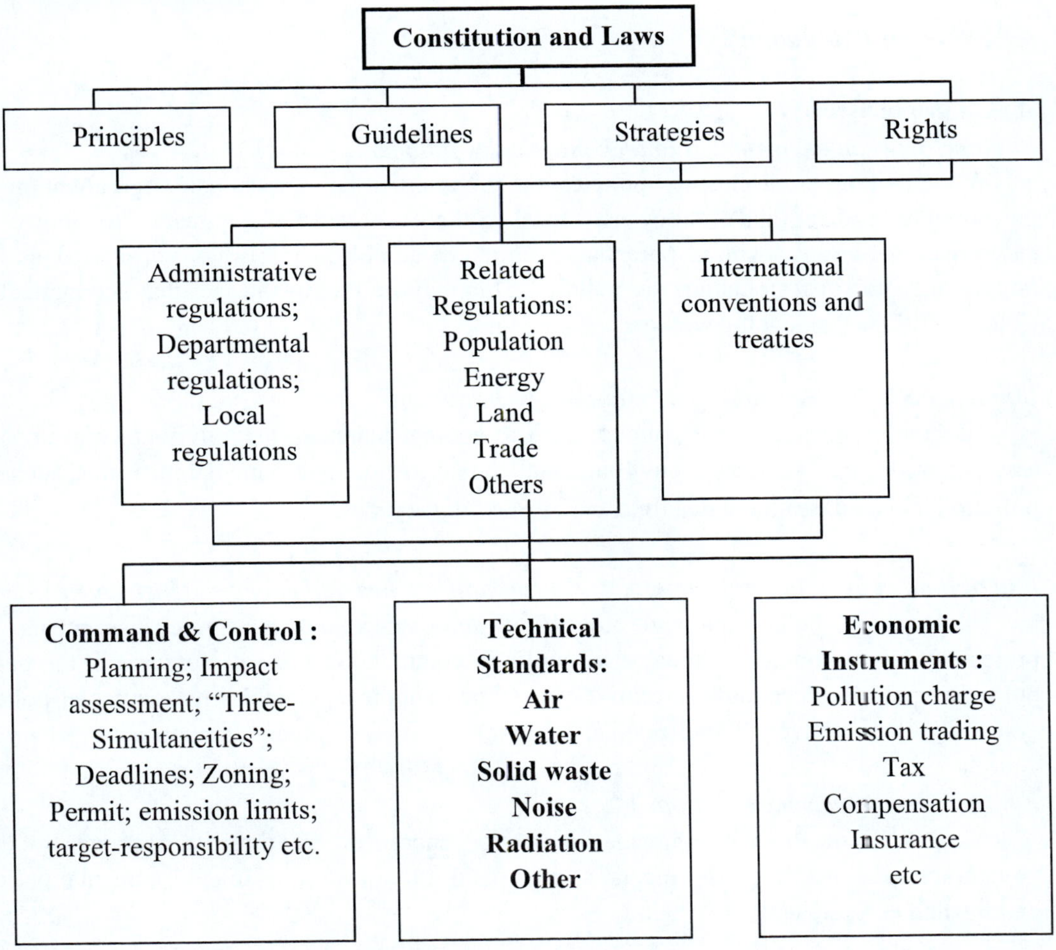

Figure 20. China's Environmental Policy Framework.

3.1.1 Policy Framework

This section intends to provide an evaluation of all countermeasures to environmental problems that China has employed in the past 30 years, which include principles and strategies that CCP and the government established, laws and regulations that were enacted by Chinese regulatory authorities, as well as policies, standards and decisions that the government authorities used to manage the environment. Figure 20 gives an overall framework of the countermeasures to environmental issues in China. The whole system can be divided into 4 layers: the first one includes constitutions, environmental laws and national fundamental policies related to environmental protection; the second layer includes the basic principles and strategies for environmental protection, such as “prevention first and joint use of prevention and treatment”, “polluters treat”, sustainable development strategy, and building environment-friendly society, etc.; the third one includes environmental regulations at different levels and with different government departments; and the fourth one includes specific environmental policies, such as command and control policies, economic policies and technical standards, etc.

3.1.2 Overall Evaluation

(1) Characteristics

Balance between environment and economic development.

While making environmental policies, the government has always paid high attention to the possible burdens that a policy may have to the economic development. The ability of enterprises to comply with a potential environmental policy is always considered as an important factor in formulating the policies. This may be due to the fact that economically China is still far from well developed.

Balance between prevention and end-of-pipe treatment.

There are both preventive policies, such as zoning, planning, and environmental impact assessment, for new sources of environmental contamination, and remedial measures, such as pollution levy and treatment deadline, etc., for exiting sources.

Use of both conventional command & control approaches and economic instruments.

While using both command & control approaches and economic instruments in protecting its environment, China relied more on command & control measures in the past, but more and more economic incentives have been adopted. China’s environmental policy system appears relatively complicated as sometimes several policy instruments act together.

Emphases on government actions.

Almost all of the environmental protection measures are directly operated by the government sectors. Nongovernmental sectors plaid little or no roles in environmental policy making and enforcement.

(2) Future Challenges

Some advanced concepts are well developed but not well implemented.

Recently, the Chinese government developed several new development concepts such as "scientific outlook on development", "building a resource saving and environment friendly society", "sound and rapid economic development", "new path of industrialization" and "ecological civilization." However, it is very challenging to make these concepts into action.

Coordination between different departments of government is weak.

The system requires that the department of environmental protection at all levels of the government to be responsible for overall monitoring and supervision of environmental work within other departments of the same level of the government system. However, it is very difficult for the environmental departments to do so. In addition, the requirements for local governments to be responsible for environmental quality in their regions have not been implemented and they do not pay sufficient attention to environmental protection.

Legal system is incomplete.

Firstly, some legal provisions and regulations are incomplete or outdated. Some important environmental protection work lack effective legal support. For example, ecological protection measures have not been included into the "Environmental Protection Law." In addition, pollution compensation, environmental information disclosure, environmental rights, pollution permits and cleaner production etc. have not been fully regulated.

Policy enforcement and compliance are weak.

Presently, the biggest problem with China's environmental management system may be the local enforcement and compliance. There are many reasons, including local protectionism and local competition for economic development. Enforcement and compliance are better in some provinces where economic conditions are better. This is also related to local people's environmental awareness.

3.2 Evaluation of Specific Policies

3.2.1 Strategies and Principles

Chinese government started to be aware of the importance of environmental protection in the late 1970s. In early 1980s, the Central Government formally established environmental protection to be a basic national policy and clearly recognized it to be a fundamental and long-term task. Up to today, only two policies have been formally established to be basic national policies: population control and environmental protection. In 1990s, the sustainable development strategy was vigorously discussed, and the economic development model and the production and consumption pattern were reconsidered. This line of thinking continued up to the 21st century, and "the scientific outlook on development", "a resource saving and environment friendly society", "harmony between man and nature" and "ecological civilization construction" are derived. These strategic thoughts and principles did provide guidance for various environmental protection activities, and push forward new measures for environmental protection.

3.2.2 Laws and Regulations

Up to now, China has established a relatively complete system of environmental laws and regulations, and the enforcement of laws and regulations keeps improving. However improvements in regulations and laws are clearly mandated. Examples include: current environmental laws and regulations are vacant in the areas of toxic chemicals control and soil pollution prevention and control; there are conflicts between environmental regulations and the regulations made by some other sectors; some provisions in current environmental laws and regulations are not operational. As stated earlier, the key issues with current environmental laws and regulations are still with the enforcement and compliance. Local protectionism heavily interferes with environmental law enforcement. Non-compliance can be found everywhere. Monitoring facilities are not fully in place. The cost of breaking laws and regulations is low. Individuals basically are not punished for their legal responsibilities.

3.2.3 Major Command & Control Measures

(1) Environmental Planning

China's Five Year Plan System for Environmental Protection started in late 1970s. After dozens of years of development, procedural laws and regulations for environmental planning management have been established and gradually environmental plans become standardized and incorporated into the national economic and social development plans. At present, China's environmental planning has formed a relatively complete system. In the framework of five year plan for national economic development, both the central government and local governments have made a series of environmental protection plans.

Vertically, environmental protection plans include country level plans (e.g., National Eleventh Five Year Plan for Environmental Protection), regional level plans (e.g., acid rain control plan), basin level plans (e.g. pollution prevention and control plans for the three rivers and three lakes), province level plans, city level plans and county level plans. Horizontally, special plans are made for different types of media, (e.g., special plan for water environmental protection, special plan for atmospheric environmental protection), and for environmental governing capabilities (e.g., environmental supervision ability building plan, national scientific and technological development plan for environmental protection, national environmental technological management system construction plan, environmental protection standard plan and environmental laws and regulations construction plan). Provincial and municipal level plans include some comprehensive plans such as ecological city plan. In addition, environmental protection authorities and other departments sometimes jointly prepare environment related plans; for example, environmental infrastructure construction plans, including urban sewage treatment and reutilization facility construction, construction of sound treatment facilities for urban domestic wastes etc.

China's environmental protection plans plaid very positive roles in promoting harmony between environment and socio-economic development, ensuring the incorporation of environmental protection activities into national economic and social development plan and guiding various environmental protection activities. However, environmental protection plans face problems such as incompleteness and insufficient implementation.

(2) Environmental Impact Assessment and "Three Simultaneities"

Environmental impact assessment (EIA) and the "Three simultaneities" [9] are two critical measures to implement the policy of "prevention first" for environmental protection and play an important role in environmental pollution prevention and control. The "Environmental Protection Law (Trial)" issued in 1979 required that environmental impact assessment and the "Three simultaneities" should be carried out for construction projects. The "Environmental management procedures for construction projects," issued in 1981 specified in detail the requirements of environmental impact assessment and the "Three simultaneities." The "Marine Environment Protection Law", "Law of Air Pollution Prevention and Control" and "Law of Water Pollution Prevention and Control" issued afterwards made corresponding detailed provisions for environmental impact assessment and the "Three simultaneities" system. The "Management method for environmental protection of construction projects" issued in 1986 and the "Regulations on the Administration of Construction Project Environmental Protection" formulated in 1997 made more specific provisions for environmental impact assessment and the "Three simultaneities" system. On October 28, 2002, the Thirtieth Session of the Ninth National People's Congress passed the "Law of Environmental Impact Assessment" which was in effect on September 1, 2003. This special law for standardizing environmental impact assessment system made specific provisions for the content and procedure of environmental impact assessment for plan and construction projects as well as corresponding legal responsibilities.

The practices show that the environmental impact assessment is a powerful environmental management system that prevents and controls environmental pollution and ecological damage from the sources and has irreplaceable important roles for promoting harmonious economic development with environment. After approximately 30 years of development, China's environmental impact assessment system has taken shape. Under the huge pressure of rapid economic development, environmental protection departments at all levels examine, approve and control total pollutants emission through environmental impact assessment for construction projects so that the growth rate of pollution emission is much lower than that of economic production. In addition, the requirement of environmental impact assessment for development programs can provide concrete institutional guarantee and make environmental protection mainstreaming into economic decision-making process.

However, there are some weaknesses associated with the environmental impact assessment system and the "three simultaneities":

- The legal requirement of environmental impact assessment for development programs, or strategic environmental assessment (SEA), has not been fully implemented. Many departments and local governments have not truly incorporated SEA into their decision making process. About one third of provinces and regions in the country have just started the work of SEA.

---

[9] The "three simultaneities" system means that the pollution treatment facilities for new construction, reconstruction and expansion projects, technical innovation projects and regional development and construction projects must be designed, constructed and put into operation simultaneously with the principal part of the projects.

- Sometimes EIA is disengaged from subsequent supervisions and the "three simultaneities" enforcement is slack. The authorities in most areas emphasize on the examination and approval of construction projects but pay less attention to the follow-up inspection and enforcement.

- Some basic recording work is not carried out properly. EIA related information and data have not been properly recorded, collected, examined and followed up, which are very important to ensure the effectiveness of EIA in practice.

- Information disclosure and public participation in EIA are not sufficient. The concept of public participation was introduced into EIA long time ago. The "Environmental Protection Law" issued in 1989 granted the right to the public to participate in environmental management and the "Environmental Impact Assessment Law" issued in 2003 specified the principles for public participation. The "Interim Measures on Public Participation in Environmental Impact Assessment" issued in 2006 institutionally incorporated public participation into EIA work. However, public participation has not been properly practiced. The requirements for public participation as defined in some relevant policy documents are technically not operational. Some project managers have antagonistic feelings towards public participation and want to avoid potential adverse effects of public participation on their project design and implementation.

(3) Total Pollution Load Control

China's environmental management started from controlling the concentration of pollutants discharged from pollution sources. It is gradually understood that purely controlling discharge concentration of pollution can not achieve the goal of environmental quality, and attention should also be paid to the total loads of different pollutants. In March 1988, the State Environmental Protection Agency issued the "Interim Measures for Management of Water Pollution Discharge Permits" with total load control as the core measure and started pilot programs of discharge permits. The total load control, as a major move in China's environmental management, appeared in the "Outline of Ninth Five Year Plan for National Economic and Social Development and 2010 Long Term Objective" passed in the National People's Congress in 1996. The former State Environmental Protection Agency specifically prepared the "Plan for Total Load Control of Major Pollutants in the Country in the Ninth Five Year Plan Period" in order to implement the Ninth Five Year Plan for Environmental Protection.

After the Ninth Five Year Plan period, the implementation of total load control has become an important item of China's environmental protection work and the implementation is enhanced in all aspects of water environmental management, which has effectively inhibited the continuous deterioration trend of water environmental quality in China and improved and enhanced environmental quality in major regions and basins. With the comprehensive development of total load control measures for water pollution in China, some problems have also emerged. Regulations are incomplete; some requirements are not operational, and monitoring capacity is not sufficient. There is a need to have a comprehensive procedure and an action plan to regulate and implement the total load control.

(4) Pollution Treatment Deadline

Pollution treatment deadline[10] is an administrative order which is the most restrictive policy measure that China adopts for environmental protection. In the past 30 years, China has implemented such polices in the following formats: company or project shut down, temporary suspension of operation, business merge with other companies, or shift to different production lines. These orders are executed before prescribed deadlines. The implementation of such administrative orders indeed has resolved some serious environmental problems in most cases. But in some places, for various reasons, the system is not fully implemented. Some orders are distorted and become a mere formality and therefore generate little impacts.

(5) Pollution Discharge Permit System

Pollution discharge permit system has been in trial for many years in China. The former State Environmental Protection Administration issued the "Interim Measures on the Administration of Water Pollution Discharge Permit" in March 1988, which specified that on the basis of reporting and registration, water pollutants permit system was implemented for the major pollution sources and major pollutants by stages and by groups. Since the end of 1990s, the cities and enterprises that implement the system of reporting and registration for discharge and the discharge permit system have been increasing year after year in the country. According to statistics, by the end of 2005, 176,733 enterprises had been granted discharge permits; 427 cities had implemented the water pollutants discharge permit system and 504 cities for atmospheric pollutants emission. In order to strengthen environmental pollution prevention and control in focus areas, China has proposed to practice "Discharge Permit" or "Temporary Discharge Permit" for major pollution. The practices show that implementing discharge permit system is positive for carrying out pollution abatement, standardizing environmental behaviors of pollution discharge organizations, completing environmental protection plans and basin water pollution prevention and control plans, and improving regional and basin environmental quality. Discharge permit system has apparent environmental, economic and social benefits and is an environmental management system that tallies with China's national condition. However, enforcement is again a major issue, beside the issue of issuing permits. Violation of permit should be severely punished, which is normally difficult.

3.2.4 Major Economic Policies for Environment

(1) Pollution Levy

Pollution levy, or emission charge, policy is an environmental economic policy implemented starting in 1982, the earliest in China. The legal basis of the levy system was established in the "Environmental Protection Law" in 1989 and in the subsequent laws

10 The system of pollution treatment within a prescribed time limit means that the legal organs make decisions on the existing pollution sources that damage the environment and forcibly order them to complete treatment and reach the specified requirements within a prescribed time limit. The decision making rights for pollution treatment within a prescribed time limit are given by the people's governments above county level and the scope of pollution treatment within a prescribed time limit can be divided into: (1) regional treatment, refers to the treatment of a region or a water area within a prescribed time limit; (2) industrial treatment within a prescribed time limit, refers to the industrial treatment of a pollutant in an industry within a prescribed time limit; (3) enterprise treatment within a prescribed time limit, refers to the treatment of overlimit pollution discharge from an enterprise within a prescribed time limit.

concerning air, water, waste and noise pollution. In 1982, the "Interim Measures for Collection of Waste Discharge Fees" was issued, and in 2003, the State Council formally issued the "Regulations on Collection and Use of Waste Discharge Fees." The levy system was continuously improving. At present, China has formulated more than 5 categories of 100 emission charge standards for wastewater, waste gases, waste residues, noise and radioactivity, and the levy system has been implemented in all provinces, cities and counties in the country. Practice shows that implementation of the levy policies effectively improves local environmental supervision and management abilities and raises pollution abatement funds, and at the same time, improves enterprises' and public environmental protection awareness.

China's pollution levy system now is not perfect yet. The charge rate is still too low, so that there are no strong incentives generated for most of the polluters. Pollutant discharge data is lack of accuracy. The pollution emission data is mostly based on self-reporting and follow-up verification. Monitoring capacity is not sufficient.

(2) User Charge

Presently, China's user charge policies mainly include wastewater treatment fee and garbage treatment fee. The Ministry of Finance, State Planning Commission, Ministry of Construction and former State Environmental Protection Administration jointly issued the "Notification on Several Issues Concerning Pilot Urban Wastewater Treatment Fees" on June 4, 1997. All areas adopt different fee standards, depending on local conditions. Garbage treatment fees were initiated in 2002 when four national ministries and commissions issued the "Notification on Implementing the Charge System for Disposal of Urban Domestic Refuses and Promoting the Industrialization of Refuse Disposal".

Wastewater treatment fees are collected from enterprises, institutions and residents based on their water consumption and the fees are paid to wastewater treatment plants for operation and management. Residents' wastewater charge standard now is about 0.2~1.0Yuan/ton water and the charges are collected together with water fees. In general, China's wastewater treatment charges are not satisfactorily implemented; the complaints include low coverage of area, low charge rate, partial payments, administrative nature, and unreasonable measurement method, etc.

The garbage treatment fee system has not been fully established. By the end of 2005, of the 661 cities in the country, 400 had not started to collect domestic refuse treatment fees, which is about 60% of the total number of cities. Many cities, especially the cities located in the central west, have not established their charge systems. For those cities which collect garbage treatment fees, the charge rates are too low to cover the treatment costs. It is also very difficult to collect the charges.

(3) Pollution Permit Trading

Pollution discharge permit trading is still of exploratory in China. The former State Environmental Protection Administration started piloting permit trading in 16 cities, including air pollution emission permit trading experiments in Liuzhou, Baotou and Kaiyuan in 1991. With assistance of Asian Development Bank and cooperation of related research institutions, Taiyuan City of Shanxi Province tested sulfur dioxide emission trading in 2002. In March of the same year, the State Environmental Protection Administration carried out experiments on sulfur dioxide emission permit trade in 7 provinces and cities including Shandong, Shanxi,

Jiangsu, Henan, Shanghai, Tianjin and Liuzhou. These experiments provided knowledge on the conditions and mechanism for pollution permit trading. The issues that have to be solved before a formal policy can be issued include: legal foundation, total limit of pollution, permit distribution, monitoring and enforcement, etc.

(4) Environmental Tax

Presently, China's environment-related taxes include consumption tax, resource tax, vehicle and vessel usage tax, vehicle purchase tax, urban maintenance and construction tax, city and town land use tax and farmland use tax etc. In recent years, measures are added to exert tax punishment by increasing tax rate for some products that cause environmental pollution, and tax credits are given to environment friendly products.

The extent of tax greening in China as a whole is still very low. First, China has not built a tax system aiming at environmental protection and has no special taxes on the behaviors or products that pollute or damage environment. Secondly, although tax revenue from environment-related taxes such as resource tax and consumption tax is 10.7% of total taxes, these taxes are not environmental taxes in real sense, because the setup of these taxes seldom considers environmental factors and has little incentive for protecting the environment.

(5) Ecological Compensation, Green Credit and Green Security

Three new economic instruments for environmental protection are currently under study and experiment in China. One is ecological compensation policy which intends to charge those activities that use or damage ecological functions of the environment such as resource extraction, and to compensate those activities that preserve or recover ecological functions of the environment such as reforestation. In September 2007, the State Environmental Protection Administration issued a "Guidance on Piloting Ecological Compensation," in order to explore compensation models, standards and policies and establish an ecological compensation mechanism.

The second one is the green credit policy. On July 12, 2007, the State Environmental Protection Administration, the People's Bank of China and China Banking Regulatory Commission jointly released a policy document "Opinions on Implementing Environmental Protection Regulations and Preventing Credit Risks" which provides principles for the banks in dealing with the environmental issues related with their lending projects

The third one is the green security policy, which, like the green credit policy and the green insurance policy, is an important integral part of policy system for a green capital market. In order to fulfill the requirements that enterprises should disclose environmental information as specified in the State Council's "Decision on Implementing the Scientific Outlook of Development and Strengthening Environmental Protection" and the requirements for strengthening environmental protection examination of listed companies in the stock market as specified in the State Council's "Notification on the Comprehensive Work Plan for Energy Saving and Pollution Reduction", in February 2008, the State Environmental Protection Administration issued the "Guidance on Strengthening Environmental Supervision for Listed Companies," which is also called "Green Security" policy, to guide the listed companies to actively fulfill their social responsibilities for environmental protection and drive the listed companies to continuously improve environmental performance. The green security policy proposes three basic institutional frameworks for environmental supervision and management of listed companies in China: (1) environmental protection examination system of listed companies; (2) environmental information disclosure system of listed companies; and (3) environmental performance evaluation system of listed companies. These three systems form a set of systematic supervision and restriction mechanism to prevent those “high pollution and high energy and resource consumption” enterprises from application for being listed in the market and refinancing after being listed.

It is too early to evaluate these three new economic measures.

3.2.5 Other Policy Instruments

Beside command & control measures and economic instruments, a set of other environmental policy instruments, mostly voluntary approaches, are developed and implemented. Examples include cleaner production certificate, environmental information disclosure, ISO 14000, and environmental labeling certification, etc.

(1) Environmental Information Disclosure

Environmental information disclosure, community pressure and participation of nongovernmental organizations are playing important roles in environmental management in a number of regions. Information disclosure enables consumers to choose less environment-harmful products and services. Public participation can help realize information transfer, information exchange and supervision. Environmental information disclosure and public participation are more and more valued in China. Environmental information disclosure policies adopted in China so far are shown in Table 5.

(2) Environmental Management System

In 1997 China State Bureau of Quality and Technical Supervision introduced ISO14000 system as a national standard. It introduced ISO14001- certified national approval system, and at the same time, the certification inspection system for registration of national agencies and auditors. In 2004, the number of certified companies in China exceeded 8,000, being the second largest in the world (next to Japan). This growth was driven by the tax and fee reduction measures that Chinese Government implemented to encourage companies to carry out certification.

(3) Cleaner Production Audit

Since the concept of “cleaner production” (CP) was introduced in 1980s, former State Environmental Protection Administration had focused on project demonstration, training, capacity building and policy recommendations at factory level. In 1997, the State Environmental Protection Administration issued the “Proposal for Promoting Cleaner Production in China” and required the local environmental protection authorities to incorporate cleaner production into their environmental management policies.

Up to now, more than 5,000 enterprises, in chemical industry, light industry, electric power, coal, machinery and building material industries, have passed cleaner production audit, and over 12,000 enterprises have been certified for ISO14000 environmental management system. More than 800 enterprises and 18,000 types of products at various sizes have passed environmental labeling certification. 21 cleaner production centers have been established in China, including 1 national level center, 4 industrial centers (petrochemical, chemical, metallurgical and aviation industries) and 16 local centers. Besides these cleaner production centers, many organizations in the country have established internal cleaner production groups or formulated cleaner production plans. Meanwhile, Chinese Government provides favorable legal conditions for cleaner production. For example, the "Cleaner Production Promotion Law" was promulgated on January 1, 2003, and it will be a propeller to further disseminate cleaner production.

Table 5. Environmental Information Disclosure and Public Participation Policies Adopted by China

Specific policies	Implementing departments	Starting time	Objects	Scope of implementation
Enterprises' environmental performance disclosure	Environmental protection authorities etc.	Piloting stage	Enterprises in pilot cities	Pilot cities
Environmental quality bulletin and weekly report	Environmental protection authorities etc.	1997	Main cities	Countrywide
Quantitative evaluation of urban environmental governance	Environmental protection authorities etc.	1984	Main cities	Countrywide
Environmental labeling	Environmental protection authorities etc.	1994	Enterprises applying for labeling	Countrywide
ISO14000 certification	Environmental protection authorities etc.	1996	Cities, regions and enterprises	Countrywide
Direct public participation in supervision	Environmental protection authorities		Public related objects	Countrywide
Supervision by media (e.g. focus report etc.)	Government and news agencies		Environmental protection-related objects	Countrywide
People's congress	NPC Standing Committee	1949	NPC deputy	Countrywide
Political consultative conference	Political Consultative Conference	1949	CPPCC members	Countrywide
Public hearing	Government		Citizens or mass representatives	Countrywide
Complaint letters and visits	Government and environmental protection authorities		Citizens	Countrywide
Environment-related award selection	Government and environmental protection authorities		Officials or mass representatives	Countrywide

Research shows that cleaner production work increases production efficiency of Chinese enterprises while reducing pollution. The work reduces pollution discharge by 20% and generates 5 million Yuan economic return each year.

(4) Eco-labeling product

Eco-labeling is an important indicator of green procurement for Chinese government, and is a wind vane for China's green products. It is leading Chinese Government and enterprises progressively to a green path. Up to now, environmental labeling certification has been carried out for 56 categories of products in household appliances, convenience goods, textile products and architectural decorative material fields. More than 21,000 types of products produced by over 1,100 enterprises have obtained environmental labels and the production value has exceeded 90 billion Yuan. The "Government Procurement List of Eco-Labeling Products" issued in 2006 indicates that related Chinese departments have formally initiated the implementation of government green procurement system in China.

4. Lessons Learned

Strong efforts and significant achievements have been observed in the work of environmental protection in China, especially in the past 10 years. The environmental institutions have been continuously improving; the investment in environmental protection has kept increasing; and the pollution intensity has kept reducing. It is fair to say that, without those efforts and achievements, the environmental quality in China could have been much worse than that of today. It should also be fair to say that the environmental efforts and achievements in China in the past were not enough, as witnessed by the unbearable pollution levels in the air, water and land, by the worrisome ecological degradations, by the economic damages and the social conflicts that the environmental problems caused to the current generation, as well as by the potential constraints that the current environmental issues post to the economic development in the future.

Major issues associated with the current environmental protection system in China rest mostly in the enforcement. As discussed before, those rather advanced concepts and policy instruments for environmental protection, such as mainstreaming environmental protection (as a fundamental national policy), integration of environment and development (EIA and three synchronizations), polluters' pay principle (pollution levy system), sustainable development, cleaner production, circular economy, information disclosure, etc., have been developed and adopted in China. A comprehensive legal system and organizational structure have been set up. However, those legal and regulatory instruments are found not well implemented, and the compliance has lagged far behind. Violations of regulations are found everywhere.

The key reasons for the weak enforcement and low compliance in China may be due to its unique top-down political structure – the general public, who is at the bottom and is the victim of the environmental problems, does not have a strong present in the work of environmental protection. Local environmental enforcers look up to their superiors within the government bodies such as mayors and county governors for discretion of regulatory enforcements, much more than up to the legal requirements or public interests. While there are strong private incentives for those local government leaders to make short-term economic growth the first priority in their personal political and economic agenda, the forces to hold the local government leaders to be responsible for the interests of the general public and the future generation are generally very weak. No strong political signals for environmental protection come from the central government. No strong political pressure can be generated from the public. But the economic incentives provided by the developers and the business people have been very strong. Therefore, the environmental regulations in China cannot be well enforced by the local authorities, and an efficient balance between economic growth and environmental protection cannot be achieved.

So, the first lesson that one may learn from the Chinese environmental protection experiences may be that a strong participation of the general public in the environmental agenda and/or a strong political will of the central government for environmental protection are necessary in order to have an efficient balance between economic growth and environmental protection, especially when the desire for economic growth is very strong and the political system is not well advanced. It is worth to note that, as pointed out before, environmental information disclosure and public participation in environmental protection have been recognized in China's environmental protection system and progress has been made in these areas even though the impacts have been marginal so far, and the Ministry of Environmental Protection has started strong efforts in promoting local enforcement and compliance.

While weak enforcement and low compliance are the major negative phenomena in China's environmental protection system, there are a number of potentially positive experiences, even though not perfect, which other countries may learn from the Chinese system:

1) The environmental zoning policy, which classifies land and water systems into several types of environmental zones, where polluting activities may be prohibited, such as in the vicinities of residential areas or drinking water sources, or be allowed, such as in the industrial parks where pollution receives centralized treatment, should be a positive practice. While no systematic research has been conducted on the zoning policy that China has adopted, it is believed that this policy can reduce the potential environmental damage and increase the pollution abatement efficiency.

2) The requirements of environmental impact assessment as well as the three-synchronization, which have been widely exercised for new development projects in China for a quite long time, are also expected to have helped to reduce environmental deterioration to some extent, even though there are rooms for improvement in the design and implementation of the two policies, as discussed before.

3) The pollution levy system is also proven an effective policy instrument[11] which can provide incentives for the polluters to reduce pollution discharge and help raise funds for environmental work, even though there are issues such as low charge rate, negotiation in fee collection[12], etc. that should be addressed in the future.

4) The approach of performance rating and public disclosure, which has been recently adopted by several provinces in China, is also expected to be an effective pollution control strategy that other countries can adopt. This approach first rates environmental performance

---

[11] See Wang and Wheeler (2003 and 2005).

[12] See Wang et al (2003).

of polluters into several (five or three) categories with colors of green, blue, yellow, red and black, from the best to the worst, and then discloses the ratings to the public. Markets and communities can provide incentives for the polluters to improve their environmental performance[13].

5) Recently the community environmental dialogue approach has been successfully tried in dozens of Chinese municipalities. Representatives of government authorities, polluters and citizens are brought together to exchange information and views about some specific environmental issues that a community is facing, discuss about how to improve the environmental status, and reach agreements on the next steps of action. A survey study has shown this dialogue approach can be cost-effective[14].

6) Some other policy measures such as administrative orders for shut-down or renovation of polluting companies can also be effective in China, but they may be less cost-effective and China-unique and may not be recommended to other countries.

REFERENCES

Chinese Academy for Environmental Planning, "*Study Report for China Environmental Economic Accounting*" (2004-2006). (in Chinese)

Dasgupta S., Laplante, B., Manmingi, N., and Wang, Hua, "Inspections, Pollution Prices, and Environmental Performance: Evidence from China," *Ecological Economics*, 2001.

Dean, Judith, Lovely, Mary, and Wang, Hua, "Foreign Direct Investment and Pollution Havens: Evaluating the Evidence from China," *Journal of Development Economics*, 2009.

Di, Wu , Chongyou, Wu, Analysis of the Evolution of Chinese Environmental Policies since the Founding of New China[J]. *Journal of Dalian University of Technology*, December 2006. (in Chinese)

Haozhan, Tan, Review, *Present Situation and Prospect of China's Environmental Polices*[J]. Journal of the Party School of Nanning Committee of the CPC, December 2006. (in Chinese)

Hebei Province Hengshui City Environmental Protection Bureau. *Let the System of Pollution Treatment within a Prescribed Time Limit Truly Exert Actual Effect*[J], November 2004. (in Chinese)

Jinnan, Wang, Hong, Long, Chazhong, Ge, *Evaluation of Impact of Environmental Policies on Society Economy[R],* Chinese Research Academy of Environmental Sciences, 2005. (in Chinese)

Kang Li, *Environmental Policy*[M], Tsinghua University Press, May 2000. (in Chinese)

Ministry of Environmental Protection, "*Ambient Air Quality Standard*" (in Chinese) (GB3095-1996).

Ministry of Environmental Protection, "*China Environmental Quality Bulletin*" (in Chinese). (1984-2007).

Ministry of Environmental Protection, "*China's Statistical Yearbook on Environment*" (in Chinese) (1996-2007).

[13] See Wang et al (2004).

[14] See Wang (2009).

Ministry of Environmental Protection, "*China's Statistical Yearbook on Environment*" (in Chinese) (1996-2007).

Ministry of Environmental Protection, "Environmental Quality Standard for Surface Water", (in Chinese) (GB3838-2002).

National Bureau of Statistics of China, State Environmental Protection Administration, "*China's Statistical Yearbook on Environment 2006*". (in Chinese)

Roumasset, James, Burnett, Kimberly and Wang, Hua, "Environmental Resources and Economic Growth in China," in Loren Brandt and Thomas G. Rawski, *China's Great Transformation: Origins, Mechanisms, and Consequences of the Post-Reform Economic Boom*, Cambridge Univ. Press, 2008.

State Council Information Office, *China White Paper on Environmental Protection*, (1996-2006), June 2006. (in Chinese)

Wang, Hua and Changhua Wu, "Environmental Institutions in China," in Urbanization, Energy, and Air Pollution in China, *The National Academies Press*, Washington, U.S. 2004.

Wang, Hua and Jin, Yanhong, "Industrial Ownership and Environmental Performance: Evidence from China," *Environmental and Resource Economics*, 2006.

Wang, Hua and Wheeler, David "Financial incentives and endogenous enforcement in China's pollution levy system," *Journal of Environmental Economics and Management*. 2005.

Wang, Hua and Wheeler, David, "Equilibrium Pollution and Economic Development in China" *Environment and Development Economics, 2003.*

Wang, Hua, "*Community Environmental Roundtable Dialogue in China*," World Bank, forthcoming.

Wang, Hua, "Pollution Regulation and Abatement Efforts: Evidence from China" *Ecological Economics, 2002.*

Wang, Hua, Bi, Jun, Wheeler, David, Wang, Jinnan, Cao, Dong, Lu, Genfa, and Wang, Yuan, "Environmental performance rating and disclosure: China's GreenWatch program," *Journal of Environmental Management*. 2004.

Wang, Hua, Dong Cao, Genfa Lu, and Jinnan Wang, "*Environmental Information Disclosure in China: Theory and Practice*," (in Chinese) *China Environment Press*. 2002

Wang, Hua, Mamingi, Nlandu, Laplante, Benoit, and Dasgupta, Susmita, "Incomplete Enforcement of Pollution Regulation: Bargaining Power of Chinese Factories" *Environmental and Resource Economics*, 2003.

Wheeler, David, Wang, Hua and Dasgupta, Susmita, "Can China Grow and Safeguard Its Environment? The Case of Industrial Pollution," in Nichloas C. Hope, Dennia Tao Yang and Mu Yang Li (eds.), *How Far Across the River? Chinese Policy Reform at the Millennium*, *Stanford University Press*, 2003.

World Bank, "*Chinese Environmental Cost Model*", 2004.

Wu, Changhua and Wang, Hua, "China: Seeking Meaningful Decentralization to Achieve Sustainability," in Albert Breton, Giorgio Brosio, Silvana Dalmazzone, and Giovanna Garrone (eds.), *Environmental Governance and Decentralization: Country Studies*, Cheltenham: Elgar, 2007.

Yanmei, Liu, *Initial Analysis of the Overall Promotion of Permit System of Pollutant Discharged in China*[J]. Yunnan Environmental Science, February 2004. (in Chinese)

Yuantai, Sun, Lili, Sun, *A Commentary of Environment Protecting Policies in Contemporary China*[J]. Decision-Making & Consultancy Newsletter, May 2006. (in Chinese)

Zhou, Guomei and Wang, Hua (Editors), *Mainstreaming Environment in China*, China Environment Press, 2008.

Zuxin, Xu, Consideration and Suggestion of Consummating of the Environmental Protection Law of China[J], (in Chinese) *Environmental Protection*, March 2007.

In: Pollution in China
Editor: Michael I. Chang

ISBN: 978-1-61122-022-3
©2011 Nova Science Publishers, Inc.

Chapter 2

ENERGY POLICY: UNDERSTANDING IMPLEMENTATION IN CHINESE FACTORIES

***Mark Yaolin Wang*[1] *and Samantha Mikus*[2]**
[1]Resource Management and Geography, University of Melbourne, Australia
[2]University of Melbourne, Australia

ABSTRACT

This study examines the status and the factors that influence how factories in Zhuji, Zhejiang negotiate China's national energy policy of reducing energy intensity by 20 percent by 2010. Using six factories as case studies, the research paper further examines the investments, modifications and motivations for change that have emerged in direct response to the national policy which aims at pollution reduction and energy reduction. Interviews conducted in the field with different company members, help to piece together a localized view of how the policy operates. Results of the field study show that ownership type effects the level of support in education and funding to reduce energy intensity in factories. Recommendations include greater education on government incentives and the fostering of stronger business networks conducive to innovation and knowledge sharing. Additionally, appropriate market-based policy mechanisms for private industry are needed to extend current implementation of energy intensity in industry.

Keywords: China, energy policy, energy intensity, Zhejiang, Zhuji, State-owned enterprises, private companies, pollution reduction

In 2005, as part of its Eleventh Five Year Plan (2006-2010) (FYP), China introduced a policy to reduce energy intensity by 20% by 2010 on 2005 levels. This was in response to a range of new challenges including resource scarcity, security of energy supply, environmental protection, climate change and pollution control. By May 2009 China had achieved a 14% reduction in energy intensity (Xinhua 2010). Chinese Premier Wen Jiabao has demanded a push to reach the full target in 2010, the final year in the Eleventh FYP (Cster 2010). As a next step, China announced in late 2009, prior to the UN Climate Summit in Copenhagen, that it would be adopting a domestically-binding goal of cutting energy intensity by 40-45 percent from 2005 levels by 2020 (Xinhua 2009).

Rapid economic growth, coupled with heavy reliance on coal, has rendered China a major contributor to global climate change. In 2008 China surpassed the US as the world's largest net emitter of CO_2 (Gregg et al 2008). A continued rapid rise in emissions would hold serious negative consequences for both China and the wider world. With China's economy predicted to continue growing rapidly for many years, effective measures for reducing energy intensity (i.e. energy used per unit of economic output) are therefore of paramount importance.

Although targets may be in place, implementation is often difficult. This research looks at opportunities for improving the implementation of China's energy intensity reduction policy. The study aims for an understanding of how national, provincial and in particular local government think about and enact energy policies within the industrial sector. Looking in detail at the policy and Zhejiang, we see how the policy is manifest in a particular province. The thesis concludes by offering recommendations for improving and strengthening the implementation of the energy intensity policy. China's administrative mechanisms for enforcing energy intensity reduction targets are enacted primarily through SOE and key targeted industries. However with burgeoning private owned enterprises (POEs) holding greater influence, new market based mechanisms need to be developed in order to reach private industry, small industry, less energy intensive industries. This will enable the energy intensity reduction policy to reach is full potential of savings and possibly achieve the aspirations of the next 5YP.

BACKGROUND

China as an environmental state has developed largely in the past thirty years. China has been an active participant in international environmental negotiations, most importantly the United Nations Framework Convention on Climate Change (UNFCCC), including the most recent Conference of the Parties in Copenhagen (COP15). China is also a signatory to the Basel Convention, Montreal Protocol and the Kyoto Protocol. While much has been written questioning China's cooperation with multilateral environmental initiatives thus far, China's general trend towards greater integration with the international community has, it is argued, been accompanied by a slow but steady increase in its commitment to conservation efforts both domestically and internationally.

Beginning in the late 1970s the Chinese central government has established an elaborate system of regulations and agencies to support environmental goals, including twenty national laws and many associated regulations (see Alford and Shen 1998). Recent years have seen China engage in particular with those multilateral initiatives relating specifically to energy: It is a member of the energy group of the Asia-Pacific Economic Cooperation (APEC), Association of Southeast Asian Nations (ASEAN) (10+3) Energy Cooperation, International Energy Forum, and Asia-Pacific Partnership for Clean Development and Climate. Nonetheless this elaborate web of environmental provisions has arguably had only a limited effect thus far. The divide between China's policy goals and reality has been highlighted

(Muldavin 2000, Ho 2001). This shows that what is needed is better implementation to achieve environmental protection and pollution control.

Some Chinese legal scholars believe that with the difficult compromise between environmental and economic interests, laws can be ambiguous and occasionally contradictory (McElroy 1998), many being compromised in implementation. It has been argued that Maoist ideals of conquering nature and maximizing productivity remain deeply embedded in the modern Chinese psyche, making a shift towards moderation and sustainability all the more challenging (Bradshaw 2007).

The Chinese government is the sole decision making body in China responsible for large-scale actions to address sustainability problems. Its unusual level of direct authority as compared with the governments of other large economies places the Chinese Government in a strong position to implement solutions to the challenges of environmental protection within its borders. China's sheer size means that any effective sustainability measures have the potential to contribute significantly to the welfare of the global environment (Grumbine 2007). However as mentioned, an implementation gap has lessened the impact of environmental policy and legislation.

Since 1978, the decentralization reforms and the reforms in planning and fiscal systems, in particular, have brought about large changes in the context of central-regional/local relationships (Morrish 1994). Beginning in 1978, a series of reforms have handed a greater role to market forces (Zhao & Zhang 1999). There have also been efforts to decentralize the previous system, and give greater autonomy and more incentives to localities. This has changed the relationship between the central and local governments (Muldavin 2000).

Since the 1980s, under Deng Xiaoping's "let someone get rich first" philosophy, (Zhao 1996), the central government has invested in its Eastern Coastal Development Strategy. This reform period has seen the birth of private industry in China and also the privatization of smaller state-owned enterprises (Ram Mohan 2004). Since decentralization, the localities have developed strategies to cope with and sometimes block the central policy (Chen 1995). Reforms that sacrifice local government interest are often viewed negatively and attacked by the local elite. The reforms have meant that the center has lost significant power to control the national economy.

Drawing on Freeman's stakeholder theory (1984), whereby an organisation will address its morals and values in managing the organisation to maximize interest of its stakeholders, it is interesting to note the change in 'stakeholders' from SOE to POE with decentralisation. It is more difficult to enact central policy through POE's.

China has many agencies that ensure compliance and interpret directives passed by higher-level authorities regarding environmental protection and pollution control work. This opens the possibility that each agency makes its own interpretation of policy (Chan et al. 1995), becoming "directives" at provincial and local government level. Increasingly, environmental strategies and the execution of specific policies occur at the local level (Banks 2001 and Ho 2000). The chains of command in the hierarchical, vertical levels of government combined with horizontal departments across government are part of the difficulty in implementation, this is known as tiáotiáo-kuàikuài 条条块块 (See Quansheng 1992 for more). The problem of tiáotiáo-kuàikuài is that it causes fragmentation and makes coordination challenging.

Given China's decentralisation and tiáotiáo-kuàikuài, it is necessary to understand the local level workings of environmental policy to achieve better implementation. China's policy to reduce energy intensity by 20% by 2010 on 2005 levels, the focus of this paper, is a recent policy and there is therefore little empirical data with which to understand its workings at a local level. This study is therefore necessary and will provide a window of increased understanding as to how industry perceives and performs their tasks in relation to the energy intensity policy.

RESEARCH METHOD

To answer the primary research question - *What factors influence how factories in Zhuji negotiate the national energy reduction target?* – the study employs an evidence-based approach based on case studies. Six factories in Zhuji, Zhejiang were selected as case studies and fieldwork carried out at the sites in January 2010. Fieldwork included a number of semi-structured interviews at each site. Using a case study analysis gives the reader a chance to understand the lines of authority within public administration and industry in Zhuji. It can also shed light on the allocation of resources and the level of policy coordination and implementation. Energy management and environmental protection are emerging areas of policy, this local perspective is aids to better understand the local experience, which gives an inside perspective of the policy relations between government and industry.

Specifically the study has been carried out in Zhejiang province. It is notable to be aware of the difference in characteristics of provinces, as State policy has a critical influence in bringing economic growth in selected regions (Morrish 1994). Zhejiang lies on the eastern coast of the People's Republic of China, and has seen huge economic growth in the past thirty years due to preferential economic policies in this area.

Zhejiang economy is grounded in manufacturing such as; textiles and construction materials, as well as food production. In recent years it has prioritized entrepreneurship of small POEs that produce bulk low cost goods. In the process it has become one of the richest provinces in China. Zhejiang also has a low number of SOE compared to POE. Due to these combined, distinct characteristics of Zhejiang, a case studies approach may uncover an unusual window of understanding of how the policy is being implemented in such as region.

Semi-structured interviews were used as the primary method. A private owner of two factories in Zhuji, agreed to assist the project with a snowball sampling technique, facilitating interviews with six local factories. Supplementary document data was also collected, in the form of energy bills. **Table 1** details the participants of this project, their role in the company, and the type of company ownership.

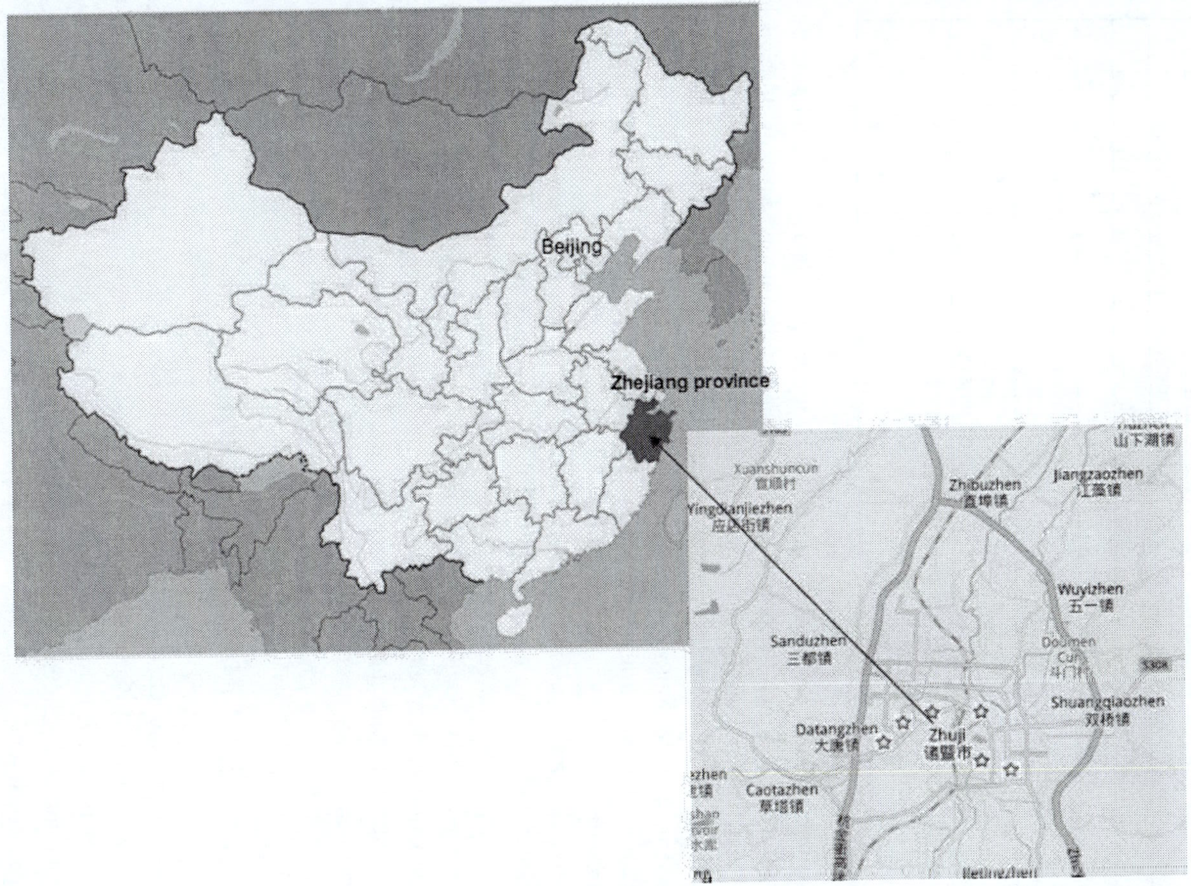

Picture 1. Above picture shows Zhejiang province in China. Lower picture shows Zhuji city and approximate location of factories.

(Source http://maps.google.com.au/)

Table 1. Participants by Role and Factory Type

Pseudonym	Main product	Role in company	Ownership
Mrs. A	Wooden doors	Owner	Private
Ms. Aa	Wooden doors	Management staff	Private
Mr. B	Silicon rod	Chief Engineer	Private
Mrs. B	Silicon rod	Owner	Private
Mr. C	Metal truck-part	Owner	Private
Mr. D	Wooden kit furniture	Co-owner	Private
Mr. E	Textile printing and dyeing	Vice General Manager	SOE
Mr. F	Electronic sock and cashmere jumper machinery	Co-owner	Private

The semi-structured interviews followed a series of broad questions that were based on a number of themes through which the practical implications of the energy intensity policy and other environmental and pollution control initiatives could be understood. To answer the research question it was necessary to explore such things as education, networking, research and development, access to funding, management structure, size, views on environment, energy and pollution. The questions were deliberately detailed to draw a larger picture from the combined answers.

The qualitative analysis is relevant for understanding how and why certain outcomes for the energy reduction policy were achieved, rather than examining just what was achieved in terms of targets. It assists in answering questions about relevance, unintended effects and the impact of the energy policy. Gathering 'thick description', as Geertz (1937) describes it, would lay a foundation of detailed accounts of a social setting which form the basis for the creation of general statements about factory culture and the significance of the energy intensity policy in peoples business lives.

Policy Background

Energy conservation and energy efficiency have been a part of China's government policy since the early 1980's. In 2004 China suffered serious shortages of all major sources of energy (Andrews-Speed 2009). In response the government stepped up its energy policy measures and made plans to embark on a program of vigorous measures to halt and reverse the rise in energy intensity.

In order to achieve these economic development aspirations in a sustainable and environmentally responsible manner, China acknowledged it needed to reduce energy intensity. The current Eleventh Five-Year Plan, 2006–2010 (National Development and Reform Commission 2007) outlined economic aims is to quadruple GDP per capita in 2020 against 2000 levels, yet this is coupled with a policy of decreasing energy intensity by 20 percent by 2010, on 2005 levels. Nonetheless the underlying message in various government policy and legal documents is that energy conservation and ecological protection are a subsidiary concern to economic growth.

Energy Intensity Explained

In aspiring for greater energy independence, mitigating climate change and reducing pollution, China has focused on reducing energy *intensity*. Energy intensity is a measure of energy use per unit of GDP. Therefore, coupled with rapid growth in GDP a policy of reduced energy intensity may not lead to a net reduction in energy use. Nonetheless, it may be argued that energy intensity is a more appropriate measure of progress in rapidly developing country. If the level energy intensity is combined with measures of, amount of renewable energy, measure of deforestation and reforestation, it may give a more in depth view and thus more accurate view of the country's progress (McMahon 2009).

Some commentators dismiss China's energy intensity target as 'business as usual' (BAU) (Levi 2009, Carraro and Tavoni 2009). However over the last five years, China has taken strong actions to address climate change, which have already lowered its CO_2 emissions. Following this, the 12th FYP targets will continue energy intensity reduction and China will need to do even more to reach its new intensity target of 40-45%. Australian economist Ross Garnaut (2009) calculates that China's dramatic policy change, of energy intensity reduction, commitment to low emissions energy and active commitment to reforestation, means a cut of 25 to 30 per cent below BAU levels of carbon emissions. Although China has hesitated to sign an internationally enforceable commitment at Copenhagen, it is stepping up commitment to environmental responsibility. Other nations, such as America are not doing enough, pledging to cut only 17 per cent from 2005 levels by 2020.

National Energy Intensity Reduction Policy in China

The 11th FYP sets a target of a 20 percent reduction in energy intensity and also incorporates additional indicators on water saving and pollution control. This was the first time that quantitative indicators for energy efficiency have been incorporated in a FYP.

The government has primarily utilized administrative and regulatory tools to achieve the 20 percent reduction in energy intensity (World Bank 2008). Administrative orders ensuring provinces and key SOEs achieve set targets have been essential to success. The government has also controlled project approval to discourage the expansion of energy intensive industries such as iron, steel, cement, aluminum, lead, paper, and chemicals (NDRC 2007).

To achieve the 20 percent energy reduction, each province has been assigned energy intensity reduction targets (**See Table 3**). It is then the role of the Governor or Mayor of that locality to ensure that target is reached. The monitoring of the policy is conducted at many levels, national, provincial, prefecture level, and county-city level (NDRC 2006). The information is fed back up the hierarchical chain of command, in order for the national government to ensure targets are being reached.

China has made significant progress toward their energy intensity reduction targets (although there has been some debate on the governments publication on official statistics:

see Howes 2010). The gains were largest in some of the most energy-intensive and polluting sectors, including power generation, steel production and mining, where the governments gave targets to individual companies and monitored them closely (World Bank 2008).

Table 2. The over-arching target of the 11th FYP is a 20% reduction in energy use per unit of GDP. In addition, the Central government has announced target levels for a number of supporting programs to achieve this. This table provides information on the primary energy savings identified for each of the programs outlined in 11th FYP Energy Intensity Reduction Targets. (Adapted from Aden 2010)

Policy/Program	**11 FYP target**
	Primary energy (MTCE) Metric Tons Carbon Equivalent
Ten Key Projects	268
Buildings Energy Efficiency	112
Top-1000 Program	130
Small Plant Closures	118
Appliance Standards	79
Other savings including provincial programs	1146
Total Primary Energy Savings	**1709**

Table 3. Provincial Energy Intensity Reduction Targets during the 11th Five Year Plan Period

Region	Base data 2005 (tce/1000 RMB)	Target for 2010 (tce/1000 RMB)	Reduction (%)	Region	Base data 2005 (tce/1000 RMB)	Target for 2010 (tce/1000 RMB)	*Reduction (%)*
Whole China	12.2		20	Henan	13.8	11.0	*20*
Beijing	8.0	6.4	20	Hubei	15.1	12.1	*20*
Tianjin	11.1	8.9	20	Hunan	14.0	11.2	*20*
Hebei	19.6	15.7	20	Guangdong	7.9	6.6	*16*
Shanxi	29.5	22.1	25	Guangxi	12.2	10.4	*15*
Inner Mongolia	24.8	18.6	25	Hainan	9.2	8.1	*12*
Liaoning	18.3	14.6	20	Chongqing	14.2	11.4	*20*
Jilin	16.5	11.6	30	Sichuan	15.3	12.2	*20*
Heilongjiang	14.6	11.7	20	Guizhou	32.5	26.0	*20*
Shanghai	8.8	7.0	20	Yunnan	17.3	14.4	*17*
Jiangsu	9.2	7.4	20	Tibet	14.5	12.8	*12*
Zhejiang	9.0	7.2	20	Shaanxi	14.8	11.8	*20*
Anhui	12.1	9.7	20	Gansu	22.6	18.1	*20*
Fujian	9.4	7.9	16	Qinghai	30.7	25.5	*17*
Jiangxi	10.6	8.5	20	Ningxia	41.4	33.1	*20*
Shandong	12.8	10.0	22	Xinjiang	21.1	16.9	*20*

Source: World Bank, 2008, *Mid-term Evaluation of China's 11th Five Year Plan*, December 18 2008, Poverty Reduction and Economic Management Unit East Asia and Pacific Region.

Note: Unit GDP Energy Intensity is based on GDP in constant prices of 2005.

All provinces in China are required to comply with the national policy of 20 percent reduction in energy intensity. Zhejiang, along with Beijing, Tianjin, Shanghai and Jiangsu provinces have been the most successful in lowering their energy intensity (Lin 2007). These areas are all wealthier, more modernized and have lower levels of heavy energy intensive industries than the national average.

Zhuji, Zhejiang province profile

The factories examined lie in Zhuji, a county city in the province of Zhejiang. Industrial output and living standards in Zhejiang have increased at a faster rate than all provinces and cities except for Guandong (Sargeson 1999). Transformed from an agriculturally based economy in 1949, Zhejiang's economy is now driven by industry, by light industrial production and processing and component assembly. This sudden change has placed pressure on basic infrastructure and raw material providers (Forster 1998). Growth in Zhejiang has been driven by the mass production of low quality goods. Reflecting back to **Table 3,** Zhejiang is placed in the top 4 of provinces with the lowest base data level for 2005, and it agreed to a 20% reduction level.

In Zhuji, where the field research was undertaken, the main focus for energy intensity reduction is on the building industry and concrete industry. In the past five years, the concrete industry has undergone reform at the direction of the Zhuji local government. The concrete industry is heavily reliant on coal boilers for energy and is considered an energy intensive industry. In Zhuji some concrete factories have been reformed others have been closed. The concrete industry still plays a role in the local area, however sock manufacturing and pearl manufacturing, are now dominant industries. China's State Council announced a 'Comprehensive Working Plan of Energy Conservation and Emission Reduction' in 2007 to accelerate closing small plants and phasing out outdated operations in 14 high energy-consumption industries, one of which was cement (State Council 2007).

Businesses may receive rebates for successfully reducing their energy intensity from either the national, provincial or local governments, depending on the size of their energy intensity reduction project. Shaoxing, a prefecture-level city in the Zhejiang, is the authority that oversees energy intensity reduction in Zhuji. The monitoring of businesses is conducted in many ways. Formally it is done through reports submitted to the government by a business, which show annual output and annual energy consumption. Informally, a network of security cameras and moles report to government about energy intensive methods, inefficient technology or wasteful practices.

As outlined above, Zhejiang's private dominated economy has low energy consumption, due to large labor-intensive activities and has high economic outputs. It is bringing heavy pressure on resources due to commodity-intensive manufacturing activities. China's growth more broadly is an increasingly unsustainable pattern of consumption and is responsible for draining large amounts resources to support changing lifestyles and as raw materials for industry. The high demand for resources by industries, rising populations and increasing per capita income is contributing to the problem.

Results

Overview of Each Factory

Case 1

Door factory is a privately owned manufacturer in wooden fire-rated doors, security doors and engineered composite doors. They were established in 1994, and have grown significantly. They have moved site locations three times in the past 10 years to accommodate the growth of the business. The company has a low energy use as much of the work requires manual labour. The company produces 300,000 pieces of solid wooden veneered door per year, and 500,000 fire rated doors per year. The main markets are domestic with some foreign exports to Australia and United Arab Emirates.

Interviewed Mrs A, the owner, and Ms A, management staff, who shared that the company was somewhat aware of the energy intensity policy and they were not actively undertaking any energy intensity reduction measures. They were not aware of any significant penalties or benefits for engaging with the policy.

Case 2

Silicon rod factory, established in 2007, is a small-sized private-owned enterprise, with 400 employees – 80 of whom work and live on site in the main industrial district of Zhuji, the rest at a plant in nearby Jiangsu province. It handles the silicon raw material treatment process, uses a Czochralski process for the production of a monocrystalline silicon rod and also handles the wafer thin slicing processing for solar cells. The silicon material handling ability amounts to 400 tons. The company uses a crystal furnace and slicer, as well as processing and inspection facilities, all of which consume a large amount of energy. At full production the company produces close to 9 million rods annually. Energy is a significant overhead cost, as much as fifteen percent of total overhead costs (According to factory owner Mrs. B). This business noted that it was significantly affected by the global financial crisis, reducing profits by 30% in 2009 on 2008 levels. The company aims to continue its growth trajectory over the coming years if it can recover to previous profit margins.

Interviewed Mr B, the chief engineer, and Mrs B the owner, who informed that the factory was aware of the energy intensity policy, but did not believe it effected their company. They were not engaged in adhering to the policy and were not regulated to reduce energy intensity. They were unaware of any support, incentives or encouragement.

Case 3

Metal truck-part company, has been operating since 2000. It specializes in the manufacturing and exporting of vehicle-fitting and mechanical components for trucks. The owner believes that they are quiet energy intensive, as production requires extreme heating and cooling. It employs and houses about 160 employees. The annual output reaches 200,000 pieces of S-camshafts, 150,000 brake shoes, 30,000 slack adjusters (manual and automatic) and 30,000 half axles. It has increased production load by approximately 15 percent each year. The biggest changes in the business since the time of first operating are related to the branding of the product. From 2000-2003 the company focused on marketing domestically and fostering good relationships with domestic markets. Since 2003 they have been able to expand their new well-known brand to other factories to manufacture parts or them. They comply with both EU and US standards so have moved to international markets rather than domestic markets.

Mr. C the owner of this company was unaware of the energy intensity reduction policy and any associated benefits or penalties. The owner had engaged in self-motivated energy efficiency measures and cost saving measures for energy overheads. As energy was a large cost in business operations, energy saving was taken seriously in order to stay competitive and minimize business costs.

Case 4

Furniture factory, a small private business, started in 1993 making leather sofas with two staff, the owners. In 1997 they diversified to making wooden kit furniture, increasing the production load, size of the factory site and increasing employee numbers to 30 people. Of these 30, all live on site. Their initial strategy was to avoid high energy and water costs; was to limit work between 7pm-8pm, which is when electricity is at its peak price. The factory has a very simplistic production line and furniture comes without varnish. The main overhead is labour.

Co-owner, Mr. D, explained that he was aware of the energy intensity policy, however due to energy overheads being only 1-2% of overall business expenditure they had not investigated further about the incentives or benefits for participation in the policy. They were not undertaking any energy intensive reduction strategies currently.

Case 5

Dye and print company, was established in 1987. It is an affiliate of the larger core enterprise *Bigger Dye, print and assorted businesses company*, which is a wholly SOE. *Dye and print company,* was listed on the Shanghai Stock Exchange (SSE) in 1997, but still to a large degree controlled to its State Owners who hold shares in the company. *Dye and print company* has 1000 employees, two thirds of whom live on site. Its markets are both domestic and international. It is capable of producing 80 million meters of high quality printing each year.

Vice General Manager, Mr. E said the printing industry was energy intensive, water intensive and highly polluting. Energy accounts for 20% of overheads for this company, he said. The biggest changes in the past ten years have been the size of the company and the increased production capacity. The machines have also been updated, now using technology from overseas, which has made the business more cost effective to do printing. Brand recognition has increased also, currently hold reputation as one of the leading textile brands in the eastern area of China.

This company was fully aware of the energy intensity policy, and had been actively engaged in business activities to follow the policy and reduce energy intensity since 2007. Motivated by the monetary incentives of the policy as well as increased training and potential of promotion, the company invested in changing technology to become less energy intensive. They were rewarded in 2009 with 2.1 million RMB for involvement with the program and reaching assigned targets.

Case 6

Sock machine and jumper machine company opened in 1999. They initially only made and assembled electronic sock machines, but have since diversified in 2007 to add computerized cashmere jumper machines, which they have found are more profitable. The company has four private owners. There are 300 workers who all live on site. Prior to 2007 the market was domestic, however now they export to 13 countries worldwide, mostly Asia Pacific region and Europe. Less than 1 percent of overheads are spent on energy, as most of the work done is computerized or assembling.

Mr. F, a Co-owner of *Sock machine and jumper machine company* said his company was not aware of details of the energy intensity reduction policy, were not actively engaging in energy intensity reduction measures, nor aware of benefits or penalties resulting from it.

Awareness of Policy and Education

The summarized table overleaf provides a broad summary of the results. This following section details education including, initial knowledge of policy, awareness of benefits and penalties, training, business network and information sharing and finally impressions of sustainability. Following this, a section on implementation, or actions to reduce energy intensity, is separated into policy related and non-policy related actions. After this, a discussion of some further observations of note, such as the role of ownership style in policy response. These are by-and-large matters which were not anticipated prior to the field research and which came to light during the interviews.

Initial Knowledge of Policy

Dye and print company was fully aware of the energy intensity reduction policy at the national level. The company first became aware of the policy when representatives from the National Government Development Department came to Zhuji to explain the policy to Zhuji County City Government officials and industry in 2007. His business was invited to attend. The meeting detailed specific requirements for targets, as well as the benefits and penalties for involvement. Following this, the mangers decided to volunteer the business to join the scheme and comply with the policy. Mr. E noted that, as he understood it:

> "Any business can choose to report voluntarily. If they want to be part of the program they need to show they have reduced energy intensity by 4% each year, report monthly, have an energy saving work unit and devise a strategy about how they will achieve energy savings. If they can achieve all of this then they may be entitled to the government bonus".

Table 4. Summary of Results from Factory Interviews

Product	Ownership	Policy knowledge	Implementation	Outcomes	Annual turnover	Energy use per month
Wooden doors	Private	Limited	None	None	26 billion RMB	9,267 kW.h (average over 1 yr)
Silicon rod	Private	Limited	Followed building energy efficiency codes when establishing new workshop	None	15 billion RMB	37,070 kW.h (average over 1 yr)
Metal truck-part	Private	None	Self motivated changes	Cost saving and energy saving	50 million RMB	28,349 kW.h (one month bill)
Wooden kit furniture	Private	Limited	Removed boiler	Received government rebate.	10 million RMB	6,273 kW.h (one month bill)
Textile printing and dyeing	SOE	Very detailed	Policy compliant	Cost saving and energy saving	5 billion RMB	N/A
Electronic sock and cashmere jumper machinery	Private	None	None	None	100 million RMB	N/A

This response was different to those received from the other five factories. All other participants interviewed either had no knowledge of the policy or very little. Most believed the national government policy to reduce energy intensity was only compulsory for certain industries such as hydropower industry, concrete industry and chemical factories. As they had not had any representative from national or local level government approach them, they were unaware of any required involvement in relation to the energy intensity reduction targets or restrictions on energy use.

From these results, it would seem that dissemination of knowledge about the policy is not happening to smaller sized operations, lower energy users, and private industry.

Awareness of Benefits and Penalties

Print and dyeing company was fully aware of potential benefits for the company, which was what swayed them to participate. Mr. E explained that normally the standard for rebate is 200 RMB per kW.h reduced. He was aware there would be an initial outlay from his company, as the company only receives a government benefit after you have proven the reduction. The money then received is tied, as in it must be spend on energy saving technology, staff training, or research and development. Among non-monetary benefits also found to be alluring was the provision of government training. This would assist in learning the necessary techniques to improve processes and also provide information about the latest technological developments. The benefits to the company he initially hoped for were greater efficiency, which in turn would save the company money. It could also make the business more competitive and the involvement was perceived to give the company greater exposure and raise its public profile. Lastly, involvement in energy intensity reduction if done well, could mean a promotion from the government.

Print and dyeing company had not ever received a fine or any penalties, but the threat of being shut down for excessive energy use was the most severe threat they were aware of. They had heard that in 2009 more than 2000 factories were shut down in Jiangsu for wasting energy. They are not ever physically inspected at the factory for their energy use, however the many surveillance cameras that monitor activities round the clock at the factories are incentive enough for the company to comply with targets.

Again, the other five factories had less awareness of the benefits and penalties. For Furniture company and Sock and jumper machine company, they had heard of the policy but because energy costs were low already, only 1-2% of overall overhead expenditure, they didn't think it was worth finding out any more information. They both remained unsure as to any benefits for complying with the policy. Furniture company and Sock and jumper machine company, they had not pursued finding out more as they considered themselves low energy users.

In regards to penalties, no company was aware of any limitations on energy usage, such as a capped amount. Silicon rod factory discussed guidelines that were needed when they initially built their workshop, following strict safety, planning and efficient building requirements. The 11th FYP, has strengthened the enforcement and monitoring for building efficiency guidelines. Implementation rates of adherence to building guidelines in 2008 showed, 98% at the design phase and 81% at construction phase (Qiu 2009). The 11th FYP's

20% energy intensity reduction target outlines the energy-saving target for the building sector is 100 Mtce in primary energy units (Aden 2010).

After the construction stage, however, they were not aware of any support, incentives or encouragement to make changes to reduce energy intensity for their own company. Yet they were aware of other penalties for not adhering to energy efficiency.

> "The multi-silicon cells are less expensive to produce than single crystal silicon cells, but are less efficient. The government does not give permission for this type of business. They would stop your business if you were using multi-silicon" (Silicon rod factory interview with engineer, Mr. B).

Furniture company and *Sock and jumper machine company* discussed random government inspections for equipment safety, fire regulations, health regulations and so on but had never been audited for energy. Nor had they heard of any other small sized companies being audited for energy usage. *Metal truck-part company* was also unaware of energy regulations, but mentioned he was monitored for pollution regularly.

Training

To learn more and keep abreast of the latest developments in energy intensity reduction, leaders from *Dyeing and printing company* are regularly invited by government to Beijing to attend professional educational sessions about energy saving within industry with central government officials. These sessions also tell businesses how to qualify for grants and subsidies through other government programs. *Dying and printing company* explained that the Ministry of Commerce sometimes meets to discuss energy and that when it does he sends a representative from the business.

> "They are then given the task of writing an essay about energy saving. Then the best one is chosen as a case study to publish in the newspaper".

He later gave me an essay that the company had written for this purpose. He explained that the meetings helped discuss and further understand the policy, but also shared field experience with other industries.

> "It is voluntary to attend, but I always find it very beneficial so I make sure I attend".

None of the other five factories interviewed engaged in any training relevant to energy use.

Business Network and Information Sharing

Business networking and information sharing is a key aspect to helping individual companies learn. It also helps improve overall success of the energy intensity reduction policy. It is also an important platform for technological innovation. The researcher asked each company how it gained new information for technological advancement or improved processes in its field.

Dyeing and printing company gained new energy intensity reduction information via networking with other businesses. According to Mr. E, as part of the energy intensity policy, the national government orchestrated formal networking meetings throughout the year and invited SOEs, POEs, foreign owned companies - both large and small energy users - to share information and skills relating to energy intensity reduction. *Dyeing and printing company* also had its own informal network of industry associates with which it met less frequently. Both of these forums provided useful information that the company utilized to make changes in the factory.

The *Silicon rod factory, Metal truck-part company, Furniture factory, Sock and jumper machine factory, and Door factory,* all did not have active formal networks with other factories or local industries designed to gain information about energy intensity reduction techniques/equipment. All noted that fierce competition deterred collaboration.

It has been well documented that, in general, the political and corporate structures in China are less conducive to collaboration and knowledge sharing than in some Western economies. Gilboy (2004) found that, "with a few exceptions, Chinese firms focus on developing privileged relations with officials in the Chinese Communist Party hierarchy, spurn horizontal association and broad networking with each other, and forgo investment in long-term technology development and diffusion". Broadly, innovation and networking in China is such that the privileged power of SOE's have grown, stifling the innovative private sector (Garnaut 2010).

Impressions of Sustainability

Although many factory owners had discussed the financial motivation and financial benefits received from energy efficiency measures, it was important to understand the factory owners' perspectives on sustainability and energy security. The relevant findings are relayed below.

The *Silicon rod factory* owner, Mrs. B., acknowledged that the product they made, silicon rods, required considerable energy to produce. However, the final product, solar batteries, were able to provide 4 or 5 times the amount of energy over their lifetime from renewable sources as what it took to produce them. In light of this the factory owner regarded the process and product as sustainable. Mrs. B's concern was that the high-energy stages were being completed in China, and then the final product - capable of providing renewable energy without emissions - was being used overseas.

The owner was not sure if China had enough raw energy to ensure a secure supply for domestic markets in the future, but believed investing in research and technology was a safety measure for an unknown future. She also believed investment in research and development of renewable energy was needed in China and in other developing nations, as future generations were likely to demand higher technology which was less polluting. In her mind, the responsibility lay squarely with the government, they being able to control domestic energy supplies, predict future demands on energy and create policy and law that would ensure secure supply in future.

Dyeing and Printing Company felt the national policy of 20% reduction in energy intensity had directly shaped the company's direction and focus, Mr. E noted:

> "It has showed us the benefit of responding to criteria in government policy and being able to receive financial rewards".

Although environmental and sustainability factors entered the equation, they were less of a priority compared with economic factors.

In regards to the sustainability of the product and process, he said:

> "We don't market ourselves as green, energy efficient or less polluting because we are still high-energy users, but our processes are more socially responsible now".

It seemed the energy policy was forcing him to think differently.

Based on the government's focus on energy, through policies like the 20% reduction in energy intensity policy, the vice general manager believed there must be a likelihood of future energy problems for China. In identifying this, he mooted that China would need to strengthen technological research and learn from Europe, the US and Australia to help fix future energy problems. In the future he would like the national government to take a greater responsibility over energy saving and energy security. He believes government should provide even greater incentives and tax reductions for businesses to help push them in the right direction. Echoing similar words as the *Silicon rod factory,* Mr. E said the government needs to address the fact that China is exporting so many manufactured goods overseas:

> "The power consumption and environmental pollution become problems of the Chinese people, rather than the countries we export to".

Metal truck-part company felt market pressure to change and improve to stay competitive is a driving force for this company, and energy and resource use are factored in as part of that. The owner, Mr. C., identified that the company's oil consumption is quiet large and that they are trying to innovate ways to reduce reliance on oil or find ways to reuse it.

Wooden door factory believed that there were enough resources for energy production in China for future domestic use, and that it is solely the responsibility of the government to ensure an equitable and sustained supply.

While aware of environmental damage and need to control energy use, respondents consistently expressed a belief that it is the government who is responsible for acting to ensure this is mitigated and planned for. This belief may be related to China's socialist style of government. This finding is similar to Wong's empirical study (2005), who also concluded that Chinese citizens hold a strong ''government-reliance attitude towards environmental protection''.

Observed actions to reduce energy intensity

This section details the energy intensity reduction actions taken in the factories, both in direct response to government policy targets and also from self-motivation.

Policy related

Dyeing and printing company was required to meet energy reduction standards of 4% per year, or above, to qualify for a government grants and not receive any penalty. Mr. E believed the yearly 4% target was standard across all industries in China's east coast region. Once this voluntary agreement was in place it became a requirement for them to follow and achieve the target, or incur a reprimand. The reduction is achieved by a combination of efforts: energy leadership group, reward system, education and a thermal energy re-utilization project, which are detailed below.

The company has established an energy saving and CO_2 emission reduction leadership group. It is made up of different heads of different departments within the company. Established in early 2009, when the company first heard about energy efficiency standards and the new national policy, it focuses on the company's electricity, gas, coal, water and other raw materials' consumption patterns, to find ways to reduce energy consumption and reduce CO_2 emissions. The group also learns from fieldtrips, where it absorbs useful experiences and applies them to local factory context. Prior to 2009 an informal and less standardized team was in place for efficiency and cost saving measures, but it focused less on energy.

Mr. E explained a belief that energy saving and reduction had to become part of the common working environment, part of the work culture, for it to be accepted and be successful. With many of the factory's workers being poorly educated, discussing energy saving is challenging and discussions need to be pitched at a level that workers can understand and relate to.

> "Popular films that talk about the environment in a mainstream way are very helpful " he said, "That's important to helping people who aren't highly educated grasp difficult and complex concepts".

The other important task, he noted, is to educate and skill owners in environmental awareness and energy saving.

> "They are likely to lead their workers and change habits on a daily level, so making sure they are informed and committed is a priority".

As well as providing information and some training for workers, the company has devised a reward system to encourage workers' learning on energy reduction. The company's internal incentive system coincides each month with the workers' salary. Workers can achieve monetary benefits if they contribute to energy savings during the month and this is shown on their pay slip. A ranking system is drawn up of the top ten people who have decreased energy running times. However, the ranking system is not a fixed scale. It is flexible and often up to the discretion of the team leaders. It is regularly adjusted to find out what works best for different situations and different departments. Some departments have the potential to reduce energy a lot. For others, such as a labor-intensive work team, it is more challenging. So the incentive system needs to allow for this. Furthermore, the bonus range is not very big. An average salary is 2000 RMB per month per worker, and the bonuses can be approximately ten percent of the salary. The same system is used for penalizing workers: If you are found to be wasting energy the bonus system is reversed. Mr. E did admit that sometimes, on a month-by-month basis, it is hard to discern what action specifically is saving energy. Given these shortcomings the incentive scheme was continually being refined and modified.

Chun's (2009) study based in Shaanxi reveals that training and engaging employees with environmental programs is a helpful starting point for Chinese energy companies wanting to increase commitment of employees, who may work in a poor environments with poor financial rewards. This seems to support the description of the successful energy intensity reduction bonus scheme the *Dyeing and Printing Company* has devised.

The thermal energy re-utilization project is the priority task for *Dyeing and Printing Company's* action toward energy saving and CO_2 emission reduction. The company has four heat-conducting-oil-ovens. Originally the ovens had a singular energy source, which, as thermal energy cannot move between ovens, proved inefficient. Maintenance work was also difficult. After modification, singular heating was replaced by centralized heating, which increased the efficiency. It was now able to turn off heat sources according to condition and seasonal variables. According to an essay written by the company, this project enabled annual savings of 2000 tons of coal, avoided 33 million cubic meters of gaseous waste, reduced CO_2 emission by 16.2 tonnes, avoided 3.6 tonnes of dust and saved 0.56 million kW.h of electricity.

The company also reuses the thermal energy from the gas outputs of the ovens. It has installed two steam generators that convert the gaseous output of the ovens into steam at 1.5 tonnes per hour and 10,800 tonnes annually. This has saved the company 1.6 million RMB. The company is able to capture and reuse 40%-70% of the thermal energy from the gas outputs of the fixed machines. This achieves both an energy reduction and a reduced amount of gas output. The 10 fixed machines can recapture 2.5 million kcal/hr, saving 3650 tonnes of coal per year.

The factory has installed a 100 tonne-per-day wastewater thermal exchanger to recapture the thermal energy in wastewater. 80ºC wastewater can be exchanged into 75ºC hot water, which saves up to 20 tonnes of steam per day. They have also reduced the use of condensed water, saving 430 tonnes of water consumption per day and 50 tonnes of steam per day.

Every month the company is required to report to Zhuji County City Government, showing energy use and a monthly energy "action plan". The Zhuji County City Government examines the reports given by each company, selects the best ten energy reducing companies in there area, submits these top ten to Zhejiang Government, who in turn choose the top ten energy reducing companies in the province.

The company always reaches the full 4% target claims Mr. E:

> "Even if we saved 7-8% one year we would report to government that we have saved 4%, because we wouldn't be entitled to the rebate if we saved 8% one year and 0% the next year."

This demonstrates the lack of incentive to achieve energy savings greater than the defined target.

The company began trying to meet the energy reduction target through self-financing. Initially the types of changes made were low cost and less risky ones. Yet in these initial stages the company was able to reduce energy from the factory by 10,000 kW.h. Due to this reduction, in 2008, the company received a rebate of 2.1 million RMB from the government.

The company then launched into the larger scale changes of the thermal project. The project's total investment cost the company 5.5 million RMB.

> "Now that we have received the government rebate we can change some of the other higher costing methods. The grant we received is tied, as in it must be spend on energy saving technology, staff training, or research and development," Mr. E.

With the thermal project the changes now mean the company annually saves 1.16 million kWh of electricity, 31.8 thousand tonnes of gas, 5650 tonnes of coal. This amounts to savings of 9.559 million RMB annually, which exceeds the total investment of the project. The combination of saving economically and avoiding environmental damage is a bonus for the company. They report that the changes mean 33 million cubic meters of waste gas is avoided, CO_2 emissions are reduced by 16.2 tonnes and 3.6 tonnes of dust is avoided.

Non-policy related

Although unaware of the full possibility of benefits from government, *Metal truck-part company* has been making energy and cost saving measures for the past ten years. Mr. C., compared a few figures, in 2000 his overhead costings were roughly 15% energy, 55% raw materials, 8% labour costs and 22% other costs (such as marketing). Through self-motivated changes in business practices and changes in technology, the company reduced the energy expenditure per unit of production, and reduced overhead spending on energy. In 2009 the company's overhead costings were: 8-9% energy, 55-60% raw materials, 10% labour costs, 23% other (such as investments in research and product development).

With rising energy prices expected Mr. C. knew he would need to make some changes in the company to help keep overhead costs low. He decreased energy usage and energy costs by four specific strategies: (1) Timing usage for peak and off peak periods. (From 12pm-6am the price of power is 90% cheaper, so they usually do most energy intensive processes during this time.) (2) Regularly servicing machinery, so that it is always operating at maximum efficiency. The company also regularly replaces machinery, upgrading gradually when new technology becomes available. (3) Using machinery to a higher capacity. What once took three steps to complete in some of their products, now only takes one step. (4) Standardizing the sizes of their products. This allows the company to use the same machines to produce different products. For example, the manual slack adjusters differ only in the diameter of the hole, not the size of whole tool, allowing several models to be made using the same machine.

Silicon rod factory had some energy and water saving practices, but none that stemmed from government targets. Rather these changes are self-motivated. The company recycles the cooling water for the rods, uses energy efficient lighting and reuses the silicon when possible. These measures were motivated by the potential cost savings. The company had not measured for either how much energy was saved or how much they were saving financially from these changes.

In 2007 *Furniture company* used a coal boiler for heating, which was both inefficient and potentially unsafe. Due to government pressure through taxes on having a boiler, the company removed it. The company had to self-finance the change in technology, but then later received a 5000 RMB rebate for not using a boiler. The Medium and Long-Term Plan for Energy Conservation describes Ten Key Projects to reduce energy intensity (NRDC 2004), one of those key projects is removing and retrofitting Coal-fired industrial boilers (kiln). It is estimated that the Ten Key Projects will contribute 40% of the overall national-level goal of 20% reduction in economic energy intensity (NDRC 2006).

No other factories interviewed were actively and identifiably taking energy saving measures. Existing studies (Farh et al. 2004 and Wong et al. 2006) have argued that non-SOEs tend to be less committed than SOEs to ethical and environmental citizenship. From the limited sample size, it was unclear as to whether they were ethically motivated or otherwise.

Further Observations of Note

This section will look at unanticipated themes of the role of ownership style in policy response and how the policy is working uniquely in this region.

There was a difference in ownership amongst the six cases I interviewed. The *Dyeing and printing company's* parent company was a SOE. And the *Dyeing and printing company* was the only company I interviewed responding to the energy policy. This shows the power still exercised through the State system. Since China's 'opening up' policy of the 1980s, private industries have gained more freedoms, including at times having the privilege of not having to so strictly comply with all government policy.

The government has chosen to act out the policy particularly through SOEs, as shown by the '1000 Enterprise Energy Savings Program', which commits about 1000 large state-owned enterprises to specific energy-saving targets (Price et al 2008). The government holds power over these companies and is able to ensure they comply with policy regulations. These companies then act as a showcase for the potential to improve industry energy efficiency. Because SOEs are considered important to the smooth running of society in China, they are sometimes given additional state funds in times of crisis, something a POE would not have access to. Demonstrating the success of SOEs in energy saving may not be sufficient impetus for private industry, who do not have access to extra financing.

Currently energy-pricing mechanisms fail to fully reflect the scarcity of resources, its supply and demand, and the environmental cost. In the past, China's natural resources, including coal and oil, were allocated to SOEs for development at no costs. The Chinese government has called for reform of energy pricing however reforms in this area have so far been limited (China Daily 2007). The prices of key energy products continue to be tightly regulated by the government but does not reflect scarcity.

Privately owned companies account for the majority of industry in Zhuji, and the larger Zhejiang and South East coast regions. A low level of state intervention has given market forces a greater shaping influence on the region. This new growth of wealth is rapidly changing the living standards of the community. Yet it is different to the growth in other areas, such as Shaanxi, where SOEs are the wealthy centers in the community. This unique local situation may have flow on effects to policy matters, as private industry is less regulated and controlled, it seems private industry may either escape or prolong the effects of having to comply with energy policy regulations and targets.

Conclusion

Reiterating, from the data gathered, from these individuals, at this point in time, the issues that emerged are: a gap in education from different types of ownership of energy intensity reduction policy requirements and benefits. Private industry, small and medium size businesses, as well as less energy intensive industries need to be included to ensure the policy is extended to its full potential. The case studies further revealed that local government agencies need to significantly increase efforts to engage other businesses in the policy; that greater measures (including penalties) must be put in place to enforce the policy; and both formal and informal networks need to be established to foster innovation and information sharing.

This research is not sufficiently broad to allow more general conclusions to be drawn about energy policy in China. Nonetheless, its insights may provide valuable pointers both for further research and for new measures for better implementing the policy of 20% reduction in energy intensity in the cases examined. The coal-dominated energy mix in China is predicted to remain dominant for some time, thus policy and practical action for climate change mitigation need to be broadened and increased. A target of reduced energy intensity will continue to play an important part in any overarching energy strategy and policy framework to mitigate climate change, curb pollution and protect the environment.

Improving information dissemination about the energy reduction policy is paramount to success. Strengthened monitoring and reporting would allow annual reports to clearly explain what has worked and what needs improvement in terms of the implementation of the program. Education and information sharing of energy intensity reduction issues would be further reinforced through improving the networking of local industries. For the industrial sector, energy performance targets for energy-intensive industries should be used as a tool to grow innovation (Price et al., 2003) and to increase enterprise competitiveness.

Lack of financing for potential energy intensity reduction changes is a barrier to industries' involvement. Self-financing is a challenge, and only larger, more wealthy companies have the capacity to invest. The vast majority of medium/small private enterprises in China that have larger energy intensity saving potential, still face tremendous difficulties in accessing to both government funds and state-owned banks' financial services (World Bank 2008). Additionally, SOEs who have access to extra funding may be unfairly supported. Greater rates and levels of involvement in the energy intensity reduction scheme would possibly occur if financing was available from either a lending institution specific for this purpose or from government.

The energy intensity target is creating large amounts of pressure on both industry and lower levels of government. In China command and control policies are particularly successful ways to achieve gains. Nonetheless more could be done to create a truly effective and comprehensive policy. Areas for improvement include setting technical standards, audits, community consultation, and voluntary agreements (Berrah et al 2007; Wang et al 2008).

The strong and extensive institutional system in China, mean administrative tools have been effective and produced quick results for the energy intensity reduction policy (World Bank 2008). These mechanisms are useful when applied to large SOEs, particularly where performance of executives is appraised according to assigned targets. However, these measures are less effective with POEs. In light of this, market based instruments and incentives need to utilized and expanded.

Though the energy intensity target for the year 2010 may be achieved, in the longer term greater efforts are needed to address a number of apparent systemic constraints on better environmental practice. The government's campaign is having effect in the short term. The question is whether this reduction will remain sustained. The rate of economic growth, structure of economy, energy efficiency of individual sectors of the economy and behaviour of society are all vital factors that will effect the future of China's energy trends.

ACKNOWLEDGMENTS

The authors gratefully acknowledge the assistance of Mrs Si Cai Yang, Miss Vivien Yang, Mr Rex Zhang and Dr. Simon Bradshaw for their contributions to this study.

Financial support for this study was granted by A/Prof Mark Wang, Department Resource Management and Geography, University of Melbourne.

REFERENCES

Aden, N., Fridley, D., Levine, M., Lu, H., McNeil, M., Price, L., Yining, Q., Zheng, N., Zhou, N., 2010 *Assessment of China's Energy-Saving and Emission-Reduction Accomplishments and Opportunities During the 11th Five Year Plan*, Lawrence Berkeley National Laboratory, LBNL-3385E, Berkeley, California, USA.

Alford, W.P., and Shen, Y., 1998 "The Limits of the Law in Addressing China's Environmental Dilemma," in *Energizing China: Reconciling Environmental Protection and Economic Growth*, ed. McElroy, M., Nielsen, C., and Lydon, P., Cambridge, MA: Harvard University Committee on the Environment, pp.405-30.

Andrews-Speed, P, 2009, 'China's Ongoing Energy Efficiency Drive: Origins, Progress and Prospects', *Energy Policy,* 37 (4): 1331-1344 April 2009.

Berrah, N., Feng, F., Priddle, R., and Wang, L., 2007 *Sustainable Energy in China: The Closing Window of Opportunity*, The World Bank, Washington DC.

Bradshaw, S.C., 2007 *Understanding the Roots of Our Ecological Crisis,* Thesis (Ph.D.)--University of Melbourne, School of Philosophy, 2007.

Carraro, C., and Tavoni, M., 2009, 'Looking ahead from Copenhagen: how challenging is the Chinese carbon intensity target?', *Climate Science and Policy,* December 28 2009, *http://www.climatescienceandpolicy.eu/2009/12/looking-ahead-from-copenhagen-how-challenging-is-the-chinese-carbon-intensity-target/*. Accessed: 23 May 2010.

Chan, H.S., Wong, K.C., Cheung, C., and Lo, J.M., 1995, 'The Implementation Gap in Environmental Management in China: The Case of Guangzhou, Zhengzhou, and Nanjing', *Public Administration Review* 55(4), 333–340.

Chen, Y. 1995. A realistic alternative for China's development and reform strategy: Formalized decentralization. *Journal of Contemporary China* 10: 81-92.

China Daily (News Agency), 2007, 'China may raise resource tax'(ed) Zhoudong, S., China Daily News Agency, 2007-06-15 16:55, Site: *http://www.chinadaily.com.cn/bizchina/2007-06/15/content_895520.htm.* Accessed: 29 May 2010.

Chun, R., 2009 'Ethical Values and Environmentalism in China: Comparing Employees from State-Owned and Private Firms', *Journal of Business Ethics,* 84: 341-348 Suppl. 3 FEB 2009.

Farh, J.L., Zhong, C., and Organ, D., 2004, 'Organizational Citizenship Behavior in the People's Republic of China', *Organization Science* 15(2), 241–253.

Forster, K., 1998 'The Political Economy of Post Mao Zhejiang', Chapter 3 in *Provincial Strategies of Economic Reform in Post-Mao China: Leadership, Politics, and Implementation,* Chueng, P., Chung, J., and Lin, Z., (eds), East Gate Books, New York.

Freeman, R.E 1984, *Strategic Management: A stakeholder approach.* Boston: Pitman.

Garnaut, J., 2010. 'A Bunch of Hacks', *The Age*, January 18, 2010, Site: *http://www.theage.com.au/business/a-bunch-of-hacks-20100117-meb7.html.* Accessed: 18 January 2010.

Gilboy, G., 2004 "The Myth Behind China's Miracle" *Foreign Affairs*, July/August 2004.

Gregg, J., Andres, R., and Marland, G., 2008, China: Emissions pattern of the world leader in CO2 emissions from fossil fuel consumption and cement production, *Geophysical Research Letters, 35, L08806, doi:10.1029/2007GL032887. http://europa.agu.org/ ?view=article&uri=/journals/gl/gl0808/2007GL032887/2007GL032887.xml&t=2008*,Gregg. Accessed: 23 May 2010.

Grumbine, R. E., 2007 "China's emergence and the prospects for global sustainability", *BioScience* March 2007, v57 i3, p249(7).

Ho, P., 2001, 'Greening Without Conflict? Environmentalism, NGOs and Civil Society in China', *Development & Change* 32(5), 893–921.

Howes, S., 2010 'China's energy intensity target: On-track or off?', *East Asia Forum*, March 31st, 2010, Site: *http://www.eastasiaforum.org/.* Accessed: 25 May 2010.

Information Office of the State Council of the Peoples Republic of China, 2005. *The Eleventh Five-Year Plan of the People's Republic of China for Economic and Social Development, 2006-2010*, Beijing, China, Distributed by China Publications Centre.

Site: *http://english.gov.cn/special/115y_index.htm.* Accessed: 8 August 2009.

Levi, M., 2009 'Assessing China's Carbon-Cutting Proposal', *Council on Foreign Relations*, November 30 2009, *http://www.cfr.org/publication/20862/assessing_chinas_ carboncutting_proposal.html.* Accessed: 23 May 2010.

Lin, S.F., and Zhao, D.T., 2007 'The analysis of different regional characteristics of energy intensity in China', *Proceedings of 2007 International Conference on Management Science and Engineering,* 14th Vols. 1-3 2007.

McMahon, H & Seligsohn, D, 2009, "Measuring Climate Change Progress in China" *World Resources Institute,* Site: http://www.wri.org/stories/2009/02/measuring-climate Accessed: 1 November 2009.

Morrish, M., 1994 'China's east coast leaves the west behind', *Geographical* - Campion Interactive Publishing Sep, Vol. 66 Issue 9, p46.

Muldavin, J., 2000 'The Paradoxes of Environmental Policy and Resource Management in Reform-Era China', *Economic Geography*, Vol. 76, No. 3 pp. 244-271.

National Development and Reform Commission (NDRC), 2006a National Development and Reform Commission (NDRC), Overview of the 11th Five Year Plan for National Economic and Social Development, NDRC, Beijing (2006).

National Development and Reform Commission (NDRC), 2006b Press Release: *NDRC and Other Related Ministries Issued Implementation Suggestions of Ten Key Energy-Conservation Projects During the Eleventh Five-Year Plan*, July 27, 2006. Site: *http://www.ndrc.gov.cn/xwfb/t20060727_78002.htm.* Accessed: 10 January 2010.

National Development and Reform Commission (NDRC), 2007 'China's Energy Conditions and Policies (2007-2012)', *Information Office of the State Council of the People's Republic of China*, December 2007, Site: http://en.ndrc.gov.cn/ Accessed: 26 May 2010.

Oster, S., 'China's Energy Use Gets More Intense', *The Wall Street Journal*, May 6, 2010, Site: http://wsj.com/chinarealtime/2010/05/06/china%E2%80%99s-energy-use-gets-more-intense/ Accessed: 23 May 2010.

Price, L., Wang, X., and Yun, J., 2008 *China's Top-1000 Energy-Consuming Enterprises Program: Reducing Energy Consumption of the 1000 Largest Industrial Enterprises in China*, Lawrence Berkeley National Laboratory Report (LBNL-519E), Berkeley, California, June 2008. Site: http://ies.lbl.gov/iespubs/LBNL-519E.pdf Accessed: 29 May 2010.

Quansheng, Z., 1992 'Domestic Factors of Chinese Foreign Policy: From Vertical to Horizontal Authoritarianism', *Annals of the American Academy of Political and Social Science*, Vol. 519, China's Foreign Relations pp. 158-175.

Qiu, B., 2009. "From Special Inspection to Fiscal Subsidy-- Annual Perspective of Architectural Energy-Saving in China" Site: *http://www.chinagb.net/news/fugle/20090330/46797.shtml*. Accessed: 20 April 2010.

Ram Mohan, T.,T, 2004 'Privatisation in China: Softly, Softly Does It', *Economic and Political Weekly*, Vol. 39, No. 45 pp. 4904-4909.

Sargeson, S., 1999 *Reworking China's Proletariat,* MacMillian Press, Great Britain.

State Council, 2007. State Council Document 2007 15.*Comprehensive Working Plan of Energy Conservation and Emission Reduction*. May 23, 2007.

Wang, G., Wang, N., Zhoa, T., 2008 'Analysis of Interactions Among Barriers to Energy Saving in China', *Energy Policy,* 36, pp1879-1889.

World Bank, 2008, *Mid-term Evaluation of China's 11th Five Year Plan*, December 18 2008, Poverty Reduction and Economic Management Unit East Asia and Pacific Region.

Wong, K. K., 2005, 'Greening of the Chinese Mind: Environmentalism with Chinese Characteristics', *Asian Pacific Review* 12(2), 39–57.

Wong, Y.T., Ngo, H.Y., and Wong, C.S., 2006, 'Perceived Organizational Justice, Trust, and OCB: A Study of Chinese Workers in Joint Ventures and State- Owned Enterprises', *Journal of World Business* 41(4), 344–355.

Xinhua News Agency, 2010 'Chinese vice premier calls for efforts to close old power plants', Official press agency of the government of the People's Republic of China, published 2010-05-22 00:32:28. *http://news.xinhuanet.com/english2010/china/2010-05/22/c_13309010.htm*. Accessed: May 24 2010.

Xinhua News Agency, 2009 'China to cut 40 to 45% GDP unit carbon by 2020', Official press agency of the government of the People's Republic of China, published 2009-11-26 16:41. Site: *http://www.chinadaily.com.cn/china/2009-11/26/content_9058731.htm*. Accessed: May 24 2010.

Zhao, X. B. & Zhang, L., 1999 'Decentralization Reforms and Regionalism in China: A review', *International Regional Science Review,* 1999; 22; 251.

Zhao, X., 1996 'Spatial disparities and economic development in China, 1953-92: A comparative study', *Development and Change* 27: 131-60.

In: Pollution in China
Editor: Michael I. Chang

ISBN: 978-1-61122-022-3
©2011 Nova Science Publishers, Inc.

Chapter 3

PCDD/Fs Levels and Major Emission Sources in China: A Review

Zhu Jianxin*

Research Center for Eco-Environmental Sciences, Chinese Academy of Sciences, Beijing, 100085, P. R. China

Abstract

China signed the Stockholm Convention on Persistent Organic Pollutants (POPs) on May 23, 2001 and the Chinese government is just making big efforts to phase out POPs production and consumption, eliminate POPs emissions, and dispose obsolete POPs pesticides and relevant wastes. With the requirements of convention implementation, monitoring ability and management capacity on POPs have been enhanced largely during these 10 years. Especially for the polychlorinated pibenzo-p-dioxins and dibenzofurans (PCDD/Fs), there are now more than 10 analytical laboratories established in China with high resolution gas chromatography/high resolution mass spectrometry (HRGC/HRMS) and qualified for the quantitative detection of dioxins. More and more data are published publicly, which gives us a chance to review the current PCDD/Fs levels in the ambient air, water, soil and sediment samples around China and show the trend PCDD/Fs pollution situation in these years. Furthermore, a PCDD/Fs emission analysis on the concerned industries including municipal solid waste incineration, medical waste incineration, ferrous and non-ferrous metallurgy, e-waste treatment and open burning, shows the contribution of these industries to POPs pollution in the whole country and tells the current level of pollution control technology development in China, that may be served as an academic reference for the related government departments when making a PCDD/Fs control policy in the near future.

---

* Zhu Jianxin, Corresponding author, PhD, Associate Professor, Tel: 86-10-62849488, Email: zhujx@rcees.ac.cn, supported by the National Natural Science Foundation of China (50708110) and (20977105).

1. INTRODUCTION

Polychlorinated pibenzo-p-dioxins and dibenzofurans (PCDD/Fs) were widely known by the Chinese public much later than the people in other countries, when China suspended the import and distribution of Belgium and other countries against dioxin-contaminated food in June 1999. As for the PCDD/Fs emission control, a scientific research and development project "Dioxin emission and control measures from municipal solid waste incineration facilities" was launched by the original State Environmental Protection Administration (SEPA) at nearly the same time. SEPA began the construction of a dioxin monitoring laboratory in the National Research Center for Environmental Analysis and Measurement (CNEAC) in 1999 to study dioxin sampling and analysis methods for major dioxin sources and propose control measures for dioxin pollution. Then two industrial standards "Pollution control standard for hazardous waste incineration" (GWKB 2-1999) and "Pollution control standard for municipal solid waste incineration"(GWKB 3-2000) were issued by SEPA to control the dioxin emission in the incineration process. It is the first time in China to regulate dioxin emission limitations in industrial standards for strengthening implementation of the "Solid Waste Pollution Prevention Law of China" and reducing secondary pollution of dioxins and heavy metals from waste incineration. The quality control procedures and measures for dioxin monitoring and analysis were developed by CNEAC in December 2000 and it had been the only institute responsible for the dioxin emissions monitoring from waste incineration facilities at that period. SEPA issued an industrial standard "High resolution gas chromatography/ high resolution mass spectrometry with isotope dilution for determination of polychlorinated dibenzo-dioxins and polychlorinated dibenzofurans" (HJ/T77-2001) in October 2001, which make the dioxin sampling and monitoring an annual requirement for those incineration plants in China.

Chinese signed the Stockholm Convention on Persistent Organic Pollutants in May 2001 and ratified it in August 2004, which is one of the most important catalysts to promote effectively the domestic research and development in the dioxin emission and control measures. Then the first publicly published preliminary dioxin emission inventory in China was set up by Jin et al based on the emission factor method in 2004, and the estimated number was about 7,144 ~ 13,575 g (toxic equivalent TEQ) from the 13 categories of main sources of emissions (Jin et al. 2004). China submitted its "National implementation plan for the Stockholm Convention on Persistent Organic Pollutants" to the Convention Secretariat in April 2007, which set 2004 as the base year and gave a dioxin emission inventory in China (NIP, 2007). This official inventory showed that the dioxin emission from waste incineration, ferrous & non-ferrous metallurgy, power generation and heat supply, mining, transportation, controlled combustion, chemical production and landfill accounted to about 10,237 g TEQ (shown in Table 1). In 2004, the total releases of Dioxin from all types of sources in China was 10.2 kg toxic equivalent (TEQ), 5.0 kg TEQ of which was released to air, 0.04 kg TEQ to water, 0.17 kg TEQ in products and 5.0 kg TEQ in residues (Zheng et al. 2008). Dioxin releases in iron and steel and other metal production industry were the biggest, accounting for 45.6% of the total, followed by power and heat generation and waste incineration. Releases from these three types of sources accounted for 81% of the total releases.

Table 1. Preliminary Inventory of Dioxins Releases in China, 2004

Major sources	Release (g TEQ/a)				
	Air	Water	Products	Residue	Sub-total
Waste incineration	610.47			1,147.10	1,757.57
Ferrous and non-ferrous metallurgy	2,486.20	13.50		2,167.20	4,666.90
Heat and power generation	1,304.40			588.10	1,892.50
Mineral products	413.61				413.61
Transportation	119.70				119.70
Uncontrolled combustion processes	64.00			953.00(Soil)	1,017.00
Production and use of chemicals and consumer goods	0.68	23.16	174.39	68.90	267.13
Disposal/landfill		4.53		43.20	47.73
Others	44.20			11.00	55.20
Total	5,043.26	41.19	174.39	4,978.50	10,237.34

Actually, this inventory was estimated mainly by UNEP's "Standardized Toolkit for Identification and Quantification of Dioxins and Furans" and based also on the emission factor methodology, which hardly give the PCDD/Fs levels and pollution situation because of the lack of monitoring capacity and could be improved much by using actual monitoring data from the emission source categories in China. With the requirements of convention implementation and dioxin emission control by Best Available Techniques (BAT) & Best Environmental Practice (BEP), monitoring ability and management capacity on POPs have been enhanced largely during these years. Especially for the PCDD/Fs, there are now more than 10 analytical laboratories established in China with high resolution gas chromatography/high resolution mass spectrometry (HRGC/HRMS) and qualified for the quantitative detection of dioxins. More and more data are published publicly, which gives us a chance to review the current PCDD/Fs levels in the ambient air, soil and sediment samples around China and show the trend PCDD/Fs pollution situation in these years.

2. Monitoring Capacity Building

The academic research of dioxin in China began in the late 1990s. When the International Symposium on Dioxin and Environmental Hormones was held in Beijing on October 28, 1998, only few academic laboratories had the ability for dioxin analysis. The first dioxin laboratory equipped with HRGC/HRMS was the State Key Laboratory of Freshwater Ecology & Biotechnology installed at the Institute of Wuhan Hydrobiology, Chinese Academy of Sciences. The ecotoxicology group in this lab investigated the sources and pollution situation of dioxins in Ya-Er Lake, China and explored multiple molecular ecotoxicological bio-indicators for the early warning of the presence and toxic effect of dioxins in the environment (Wu et al. 1997). A dioxins lab in Peking University was set up in 2000. The main research fields of the laboratory includes: dioxins in the soil and sediment in Beijing and dioxins exposure of the Chinese population (Chen et al. 2003). The Center for Disease Control and

Prevention of Shenzhen City was set up in 2000. Their main field of interest is PCDD/Fs and PCBs analysis in food. The dioxin laboratory in Key Laboratory of Environmental Chemistry and Ecotoxicology of RCEES was established in 2002, which emphasized POPs effect on the ecosystem and human health, the laboratory studies on its effecting mechanism, complex pollution control and remediation principle, as well as the theory and methodology of ecotoxicology with long term exposure to low concentration persistent toxic substances (Zheng et al. 2003).

CNEAC proposed to SEPA that environmental dioxin pollution control research and capacity building should be carried out in China to study the POPs situation and countermeasures in later 1990s and a dioxin analysis laboratory was built with high-resolution gas chromatography / high resolution mass spectrometry technology in Sino-Japan Friendship Center for Environmental Protection by the end of 1999 (Tian et al, 2002). Then the laboratory was refined in April 2004 and equipped with advanced equipment, such as: high resolution gas chromatography - high resolution mass spectrometer (HRGC-HRMS), gas chromatography - low resolution mass spectrometer (GC-LRMS), headspace chromatography mass spectrometer, liquid chromatography mass spectrometer, gel permeation chromatography (GPC), the sample auto-cleaning device (FMS Power-Prep TM/P2-MC), Accelerated Solvent Extractor (ASE) and ambient air dioxin sampling devices. In 2008, the refined dioxin laboratory was approved as “the State Environmental Protection Key Laboratory of Dioxin Pollution Control” to provide technical support in environmental administration for the SEPA and to provide comprehensive technical services for the community laboratory.

High speed capacity development in dioxin monitoring mainly results from that China signed "Stockholm Convention" in 2001, China's rapid economic development, big capital investment and more and more scientists involved in the POPs research. In the beginning of 2000, only several scientists were interested in POPs research because of the lack of necessary basic study and monitoring ability, when Chinese scientists published few papers in international journals and the outputs "can be ignored" in the international POPs research. To enhance the dioxin monitoring ability in provincial level, SEPA designed a plan for the construction of 7 regional dioxin monitoring centers in the “National hazardous waste and medical waste disposal facility construction program” (The Program). The Program was approved by the State Council in December 2003 and Beijing, Shenyang, Hangzhou, Guangzhou, Xi'an, Chongqing, Wuhan was just preparing their establishment of seven dioxin monitoring centers and to be responsible for the future dioxin emission monitoring tasks from the hazardous waste & medical waste disposal facilities, municipal solid waste incineration facilities and other emission sources. The State Environmental Protection Key Laboratory of Dioxin Pollution Control would be the technical supporter in the building design, technical and personnel training, standardization of laboratory management and technical training.

Rapid changes have taken place for the dioxin and POPs research in China and a series of outstanding achievements have been made from 2002 to 2005 under the strong support of the National Key Basic Research Program (973 program), National Natural Science Foundation of China (NSFC) and National High Technology Research and Development Program of China (863 plan). Especially in 2004 since the Stockholm Convention entered into force in China, high level of achievements in the studies of POPs pollution and trends, migration and degradation mechanism, biological accumulation and enlargement, toxic effects, control and elimination technology, which showed increasingly significant impact in the related international researches. By the end of 2008, four of the seven dioxin monitoring centers finished their basic construction and the operators were trained at CNEAC. The 29th International Symposium on Halogenated Persistent Organic Pollutants (Dioxin 2009) held on August 23-28, 2009 in Beijing, China offered a good opportunity and an international interdisciplinary forum for the communication between Chinese scientists and international scientists for the latest scientific advances and solution of environmental problems concerned with dioxins and other persistent organic pollutants.

The results from intercalibration study for dioxins showed that dioxin laboratories established in China were qualified for the analytical skill required to detect the dioxins among many interferences and the analytical variance with the dioxin laboratories in the western world including Europe, the US, Canada and Japan was small and quantifiable. The concentrations of PCDD/PCDF in environmental samples are normally at the ppt level (ng per kilogram or pg per gram), so intercalibration study was mainly set as a global QA/QC tool for the analysis of POPs under the Stockholm Convention. Seven dioxin laboratories from mainland of China attended in the 10th round dioxin intercalibration study supported by the UNEP- GEF Project "Assessment of Existing Capacity and Capacity Building Needs to Analyse Persistent Organic Pollutants (POPs) in Developing Countries", including: Research Center for Eco-Environmental Sciences of Chinese Academy of Sciences, Chinese Academy of Inspection and Quarantine, Dioxin Laboratory of ITPE of Zhejiang University, Guangzhou Institute of Geochemistry of Chinese Academy of Sciences, Shenzhen Center for Disease Control & Prevention, National Research Center for Environmental Analysis & Measurement (NEAC) and National Institute of Nutrition and Food Safety of Chinese Center for Disease Control and Prevention. Till now, there are 10 dioxin laboratories Listed in the UNEP Databank for Existing Dioxin Monitoring Capacity in Developing Countries, their general description and Methods Used for Identification and Quatification of POPs and Specific Matrices were shown in Table 2 and Table 3 respectively.

3. PCDD Level in Environmental Media

3.1 Sediment

More than half of the recent research on dioxins in China has focused on sediment samples. Figure 1 shows the locations where sediment samples were collected from 1994 to 2009, and Table 4 lists their PCDD/F levels. These samples were collected around the country, covering a large area and several important water systems: the Yangtze River — the largest river in China — and its catchments Poyang Lake; Dongting Lake; Ya-Er Lake; the Liao He River in the Northeast; the Pearl, Liangjiang Rivers and Hong Kong Bay in the southeast; the Nanpaiwu River and Bohai Bay in the north. The PCDD/F concentrations of the sediment samples vary from 14 to 3,289,600 pg g^{-1} (Table 4). The OCDD mass fraction was about 80%. The median PCDD/F concentration of the samples was 16.25 pg TEQ g^{-1}, which is quite close to values for lightly polluted rivers and bays in other countries, e.g., the Tama River, Tokyo Bay, and Osaka Bay in Japan (Onodera et al. 1999, Hosomi et al. 2003, Takigami et al. 2005); the Lower Great Lakes, North America (Marvin et al. 2002); and the Po River and Venice Lagoon, Italy (Frignani et al. 2001, Fattore et al. 2002).

Figure. 1. Surficial sediment sampling locations collected from 1994 to 2009 around China from the dioxin research literatures published.

Table 2. Identity and General Description of 10 Dioxin Laboratories Listed in the UNEP Databank for Existing Dioxin Monitoring Capacity in Developing Countries

No.	Name of Laboratory :	City / State :	Type of Laboratory :	Total Size (in square meters) :	Dedicated to POPs Analysis :	National and international services? :	Operational year :	Experience with POPs analysis since (starting year) :		
1	RCEES - Dioxin Laboratory, Research Center for Eco-Environmental Sciences, Chinese Academy of Sciences	Bejing	Research Center	500	300	Yes	1975	Pesticides : 1977	PCB : 1980	PCDD / PCDF : 1988
2	POPs Laboratory of Department of Environmental Science and Engineering, Tsinghua University	Beijing	Research Center	600	200	Yes	2002	Pesticides : 2001	PCB : 2001	PCDD / PCDF : 2006
3	Advanced Analytical Center of Dalian Institute of Chemical Physics, Chinese Academy of Sciences	Dalian	Research Center	600	280	Yes	1969	Pesticides : 2000	PCB : 2002	PCDD / PCDF : 2002
4	NEMC - Ningbo Environmental Monitoring Center	Ningbo	Public / Governmental	4000	910	Yes	1976	Pesticides : 1990	PCB : 2006	PCDD / PCDF : 2008
5	CNEAC - National Research Center for Environmental Analysis and Measurement	Bejing	Research Center	2000	328	Yes	1998	Pesticides : -	PCB : 1998	PCDD / PCDF : 1999

Table 2 (Continued).

No.	Name of Laboratory :	City / State :	Type of Laboratory :	Total Size (in square meters) :	Dedicated to POPs Analysis :	National and international services? :	Operational year :	Experience with POPs analysis since (starting year) :
6	SKLOG - State Key Laboratory of Organic Geochemistry	Guangzhou	Public / Governmental	3200	2000	Yes	1991	Pesticides : 1992 PCB : 1996 PCDD / PCDF : 2003
7	Dept of Environmental Biology and Molecular Ecology, Shanghai Jiaotong University	Shanghai	Research Center	320	100	Yes	1998	Pesticides : 2000 PCB : 2000 PCDD / PCDF : -
8	CCDCP - National Institute of Nutrition and Food Safety, Chinese Center for Disease Control and Prevention	Beijing	Public / Governmental	360	120	Yes	1946	Pesticides : 1971 PCB : 1971 PCDD / PCDF : 1996
9	SZCDC - Shenzhen Center for Disease Control & Prevention	Guangdong Province	Public / Governmental	300	250	Yes	1994	Pesticides : 1998 PCB : 2003 PCDD / PCDF : 2001
10	Dioxin Laboratory of Institute of Thermal Power Engineering, Zhejiang University	Hangzhou/Zhejiang Province	Research Center	700	500	Yes	1999	Pesticides : - PCB : 2005 PCDD / PCDF : 1999

Table 3. Methods Used for Identification and Quatification of POPs and Specific Matrices in the 10 Dioxin Laboratories

No.	Institutes	Media	Basic POPs								Dioxin-like POPs	PCDD/PCDF	
			Aldrin, endrin, dieldrin	Chlordane	DDT(incl. DDD/DDE)	Heptachlor	Mirex	Toxaphene	HCB	Indicator PCB	dl-PCB (TEQ)	2,3,7,8-subst. (TEQ)	Homologs
1	RCEES	Stack Emmision	EPA 8081(A)(B)		EPA 8081(A)(B)	EPA 8081(A)(B)			EPA 8081(A)(B)		EPA 1668(A)	EPA 23(A)	EPA 23(A)
		Transformer Oil								EPA 8082(A)		EPA 1613	EPA 1613
		Solid Abiotic	EPA 8081(A)(B)		EPA 8081(A)(B)	EPA 8081(A)(B)			EPA 8081(A)(B)	EPA 1668(A)	EPA 1668(A)	EPA 1613	EPA 1613
		Aqueous	EPA 8081(A)(B)		EPA 8081(A)(B)	EPA 8081(A)(B)			EPA 8081(A)(B)	EPA 8082(A)	EPA 1668(A)	EPA 1613	EPA 1613
		Biota	EPA 8081(A)(B)		EPA 8081(A)(B)	EPA 8081(A)(B)			EPA 8081(A)(B)	EPA 1668(A)	EPA 1668(A)	EPA 1613	EPA 1613
		Ambient Air										TO 9(A)	TO 9(A)
		Stack Emmision									HJ77.2-2008/ JISK 0311	HJ77.2-2008/ JISK 0311	HJ77.2-2008/ JISK 0311
		Transformer Oil								EPA 8082(A)			
2	Tsinghua	Solid Abiotic	EPA 8081(A)(B)	EPA 8081(A)(B)	EPA 8081(A)(B)	EPA 8081(A)(B)			EPA 8081(A)(B)	EPA 8082(A)	HJ77.3-2008/ HJ77.4-2008	HJ77.3-2008/ HJ77.4-2008	HJ77.3-2008/ HJ77.4-2008
		Aqueous	EPA 8081(A)(B)	EPA 8081(A)(B)	EPA 8081(A)(B)	EPA 8081(A)(B)			EPA 8081(A)(B)	EPA 8082(A)	HJ77.1-2008/ JISK 0312	HJ77.1-2008/ JISK 0312	HJ77.1-2008/ JISK 0312
		Biota											
		Ambient Air											
		Stack Emmision							EPA 8270(C)(D)	EPA 1668(A)	EPA 1668(A)	EN 1948/ EPA 1613	EN 1948/ EPA 1613
		Transformer Oil								EPA 1668(A)	EPA 1668(A)		
3	Dalian	Solid Abiotic		EPA 8081(A)(B)/ EPA 8270(C)(D)	EPA 8081(A)(B)/EPA 8270(C)(D)					EPA 1668(A)	EPA 1668(A)	EPA 1613	EPA 1613
		Aqueous								EPA 1668(A)	EPA 1668(A)	EPA 1613	EPA 1613
		Biota										EPA 1613	EPA 1613
		Ambient Air										EPA 1613	EPA 1613
		Stack Emmision	EPA 8081(A)(B)	EPA 8081(A)(B)	EPA 8081(A)(B)	EPA 8081(A)(B)			EPA 8081(A)(B)		EPA 1668(A)	EPA 23(A)	EPA 23(A)
		Transformer Oil								EPA 8082(A)			
4	Ningbo	Solid Abiotic	EPA 8081(A)(B)	EPA 8081(A)(B)	EPA 8081(A)(B)	EPA 8081(A)(B)			EPA 8081(A)(B)	EPA 1668(A)	EPA 1668(A)	EPA 1613	EPA 1613
		Aqueous	EPA 8081(A)(B)	EPA 8081(A)(B)	EPA 8081(A)(B)	EPA 8081(A)(B)			EPA 8081(A)(B)	EPA 8082(A)	EPA 1668(A)		
		Biota											
		Ambient Air										TO 9(A)	TO 9(A)
		Stack Emmision										In-house HJ-77	EPA 23(A)
		Transformer Oil								Japanese Manual		EPA 1613	
5	CNEAC	Solid Abiotic	Japanese Manual	Japanese Manual	Japanese Manual				Japanese Manual	Japanese Manual		EPA 1613	
		Aqueous	Japanese Manual	Japanese Manual	Japanese Manual				Japanese Manual	Japanese Manual		EPA 1613	
		Biota								Japanese Manual		EPA 1613	
		Ambient Air	TO 4(A)	TO 4(A)	TO 4(A)				TO 4(A)	TO 4(A)		TO 9(A)	

Table 3 (Continued)

No.	Institutes	Media	Basic POPs								Dioxin-like POPs	PCDD/PCDF	
			Aldrin, endrin, dieldrin	Chlordane	DDT(incl. DDD/DDE)	Heptachlor	Mirex	Toxaphene	HCB	Indicator PCB	dl-PCB (TEQ)	2,3,7,8-subst. (TEQ)	Homologs
6	SKLOG	Stack Emission	EPA 8081(A)(B)	EPA 8081(A)(B)	EPA 8081(A)(B)	EPA 8081(A)(B)	EPA 8081(A)(B)	EPA 8081(A)(B)	EPA 8081(A)(B)			EPA 1613	
		Solid Abiotic	EPA 8081(A)(B)	EPA 8081(A)(B)	EPA 8081(A)(B)	EPA 8081(A)(B)	EPA 8081(A)(B)	EPA 8081(A)(B)	EPA 8081(A)(B)			EPA 1613	
		Aqueous	EPA 608	EPA 608	EPA 608	EPA 608	EPA 608	EPA 608	EPA 608				
		Biota	EPA 8080(A)	EPA 8081(A)(B)	EPA 8081(A)(B)	EPA 8081(A)(B)	EPA 8081(A)(B)	EPA 8081(A)(B)	EPA 8081(A)(B)			EPA 1613	
		Ambient Air	EPA 8080(A)	EPA 8081(A)(B)	EPA 8081(A)(B)	EPA 8081(A)(B)	EPA 8081(A)(B)	EPA 8081(A)(B)	EPA 8081(A)(B)			EPA 1613	
7	Shanghai	Stack Emission											
		Solid Abiotic			EPA 8081(A)(B)				EPA 8081(A)(B)	EPA 8082(A)			
		Aqueous											
		Biota			EPA 8081(A)(B)				EPA 8081(A)(B)	EPA 8082(A)			
		Ambient Air			EPA TO 13(A)				EPA TO 13(A)	EPA TO 13(A)			
8	CCDCP	Stack Emission								EPS 1/RM/31	EPA 1668(A)	EPA 1613	
		Transformer Oil								EPS 1/RM/31	EPA 1668(A)	EPA 1613	
		Solid Abiotic	GB 5009	GB 5009	GB 5009	GB 5009	GB 5009	GB 5009	GB 5009	EPS 1/RM/31	EPA 1668(A)	EPA 1613	
		Aqueous	GB 5009	GB 5009	GB 5009	GB 5009	GB 5009	GB 5009	GB 5009	EPS 1/RM/31	EPA 1668(A)	EPA 1613	
		Biota	GB 5009	GB 5009	GB 5009	GB 5009	GB 5009	GB 5009	GB 5009	GB 5009	EPA 1668(A)	EPA 1613	
		Ambient Air								EPS 1/RM/31	EPA 1668(A)	EPA 1613	
9	SZCDC	Stack Emission								EPA 1668(A)	EPA 1668(A)	EPA 23(A)/EPA 8290(A)	EPA 23(A)/ EPA 8290(A)
		Transformer Oil											
		Solid Abiotic								EPA 1668(A)	EPA 1668(A)	EPA 1613	EPA 1613
		Aqueous			In-house				In-house	EPA 1668(A)	EPA 1668(A)	EPA 1613	EPA 1613
		Biota			In-house				In-house	EPA 1668(A)	EPA 1668(A)	EPA 1613	EPA 1613
		Ambient Air								EPA 1668(A)	EPA 1668(A)		
10	Zhejiang	Stack Emission									EPA 23(A)		EPA 23(A)
		Transformer Oil											
		Solid Abiotic								EPA 1668(A)		EPA 1613/ EPA 8280(A)/ EPA 8290(A)	EPA 1613
		Aqueous								EPA 1668(A)		EPA 1613	
		Biota								EPA 1668(A)		EPA 1613	
		Ambient Air								EPA 1668(A)		EPA TO 9(A)	

In Northeastern China, Zhang et al. studied the PCDD/Fs in sediments from Daliao River Basin(Zhang et al 2008) and the results showed that the concentrations of PCDD/F ranged from 0.28 to 29.01 pg TEQ g^{-1} dw (mean value, 7.45 pg TEQ g^{-1} dw), which was in the same range as Martens's study(18.3 pg g^{-1}) in 2000. PCDD/F pollution in sediments from the mid- and downstream sections of Hun River was found to be relatively heavy, PCDD/F contamination in most sediments of Hun River was supposed be originated from the production of organochlorine chemicals, while metal smelting was the important potential source of PCDD/F in the drainage area of Taizi River.

In Northern China, where a lots of chemical industries including the POPs pesticides had been produced for several decades, the concentration of PCDD/Fs were much higher than that in other areas. Wang, B. et al. studied the level of Dioxin-Like Compounds in Sediments from the Haihe River by CALUX Bioassay. It was found that the levels of PCDD/Fs in the sediments from the Haihe River were tens of times to hundreds of times higher than those of DL-PCBs (Wang et al. 2009). A high level of DLCs was also found at the sites near the influx into Bohai Bay. Luo et al. (2009) analyzed the characterization of PCDD/Fs in sediments of Wenyu River, Beijing(Luo et al 2009). Five samples were selected and all of them showed significant Ah-agonistic effects. The bioassay-derived 2,3,7,8-TCDD equivalents of raw extracts ranged from 8.5 to 336.0 pg g^{-1} dw. Recently, the concentrations of PCDD/Fs and PCBs in the sediments from the mainstream of Haihe River were 1.3-26 pg I-TEQ g^{-1} dw and 0.07-0. 54 pg TEQ g^{-1} dw, respectively. Heavy PCDD/Fs and PCBs pollution, with 1264 pg I-TEQ g^{-1} dw and 21 pg TEQ g^{-1} dw, was found in sediment from Dagu Drainage River (Liu et al. 2007). Hu's study in 2005 also showed high PCDD/Fs pollution from the drainage river to the Bohai Bay, e.g. the PCDD/Fs level in Nanpaiwu River was as high as 121.302–3,289,600 pg g^{-1}, and the concentration in Bohai Bay was about 447.8 pg g^{-1} at that period (Hu et al 2009).

In Southern China along the Yangtze River, Wu et al. showed that PCDD/Fs, other than HCH, were prominent pollutants in some areas of the Yangtze River after production of HCH was banned in the late 1990. Ya-Er Lake, located along the middle-lower reaches of the Yangtze River, had PCDD/F levels in the range 0.11–0.43 pg TEQ L^{-1}. Zheng's study showed that PCDD/Fs concentrations of contemporary sediment have declined from 1995 to 2004, (Zheng et al. 1997, 2001, 1003), the concentrations of PCDD/Fs in sediments in Dongting Lake were 130-891 pg I-TEQ g^{-1} in 1995 and the total I-TEQ values for sediment samples were at a ranged of 0.7-11 pg g^{-1}, with a mean value of 4.5 pg g^{-1}. Sediment samples from Meiliang Bay, Taihu Lake were assayed for AhR-mediated EROD induction using a rat hepatoma cell line (H4IIE) by Qiao et al (2006). The results showed that sediment extracts could induce significant AhR effects, and the chemical-derived TCDD equivalents (TEQ(cal)) were significantly correlated to bioassay-derived TCDD equivalents (TEQ(bio)) ($R = 0.85$, $p < 0.01$). Zhang, Q. H's study for sediments and aquatic organisms from the Taihu Lake showed that concentrations of total PCDD/Fs, PCBs and WHO-TEQ were 120.1-1315.1 pg g^{-1} dw, 889.7-29747.8 pg g^{-1} dw, and 0.83-17.72 pg TEQ g^{-1} dw, respectively in sediments (Zhang et al. 2005). The concentrations of PCDD/Fs and PCBs in the sediments were decreasing gradually along the water flow. Surface sediments were collected by Wen et al. (2008) from the Yangtze Estuary in December 2003 and November 2004, respectively. The concentrations of total PCDD/Fs measured with HRGC/HRMS were 169.83±119.63 and 0.81±0.36 pg g^{-1} TEQ dry weight (dw) in sediments, which are much closer to the Hui Y.M's results that 0.36-0.74 pg g^{-1} TEQ for Yangtze estuary (Hui et al. 2009). Li et al. (2007) studied the PCDD/Fs and dioxin like PCBs (DL-PCBs) in sediments from the Suzhou Creek and the results showed that mean concentrations of PCDDs, PCDFs, and PCBs were 478.1, 245.1, and 4727.6 pg g^{-1} dw, respectively, the WHO-TEQ concentrations of PCDD/Fs in sediments ranged from 2.90 to 13.96 pg g^{-1} dw.

Table 4. Sampling locations and PCDD/F levels of the sediment samples collected from 1994 to 2009 around China from the dioxin research literatures published

Position in China	Water system	In Fig. 1	PCDD/F concn range (pg g^{-1} I-TEQ)	Year	Reference
Northeastern	Liao River	A	18.3	1999	Martens et al. 2000
	Daliao River	A	0.88-9.49	2005	Zhang et al. 2008
	Taizi River	A	2.67-4.62	2005	Zhang et al. 2008
	Hun River	A	0.28-29.01	2005	Zhang et al. 2008
Northern	Dagu	B	425–1,456,170	2001	Luksemburg et al. 2002
	Wenyu River	B	8.5-336	2009	Luo, et al. 2009
	Hai River	B	1.3-26	2007	Liu, et al. 2007
	Dagu Drainage River	B	1,246	2007	Liu, et al. 2007
	Nanpaiwu River	B	121,302–3,289,600	2002	Hu et al. 2005
	Bohai Bay	C	447.8	2002	Hu et al. 2005
Southern	Ya-Er Lake	D	61–33837	1994	Wu et al. 1997
	Ya-Er Lake	D	2100–6620	1997	
	Yangtze River	N	751–1024	1998	Martens et al. 2000
	Yangtze River	K	23197	1999	Martens et al. 2000
	Dongting Lake	J	52617–233304	1995	
	Dongting Lake	J	641–1881	1999	Zheng et al. 2003
	Taihu Lake	K	120.1-1315.1	2008	Wen et al.2008
	Yangtze Estuary	O	0.81±0.36	2007	Li et al, 2007
	Yangtze Estuary	O	0.36-0.74	2007	Hui et al, 2009
	Hongze Lake	G	62.09	1995	
	Nanfei River	L	3944.5	1995	
	Dongting Lake	J	267.86	1995	
	Poyang Lake	M	900.49	1995	
	Yangtze River	N	162.4–462.0	1995	
Eastern	Huangpu River	O	65.29–178.29	1995	
	Han River	H	13.52	1995	
	Bengbu Harbor	F	129	1995	
	Grand Canal	F	58.5	1995	
	Fu River	I	103.9	1995	
	Huaihe River	E	515.4	1995	
	Shandong Peninsula	R	6.2–27.4	~2009	
	Yellow river estuary	R	0.11-1.01	2007	Hui et al,2009
Southeastern	Hong Kong Bay	R	314–10997	2001	Muller et al. 2002
	Liangjiang River	Q	56–743180	2002	Luksemburg et al. 2002
	Hong Kong Bay	R	9.0-18.8	2004	
	Pearl River	P	498–2590	1997	Zheng et al. 2001
	Pearl River	P	1.24-2.63	2008	
	East River	Q	0.042 to 0.45	2007	

In Eastern China, Pan, J et al. (2010) analyzed the PCBs, PCDDs and PCDFs in marine and lacustrine sediments from the Shandong Peninsula, and the results showed that the total concentrations of 2,3,7,8-PCDD/Fs ranged from 6.2 to 27.4 pg g^{-1} dw. The congener profiles of 2,3,7,8-sustituted PCDD/Fs for the sediments were generally similar for both the lakes and the coastal sea areas in Shandong Peninsula. They were characterized by high OCDD, followed by 1,2,3,4,6,7,8-HpCDD and OCDF. Hui et al. (2009) took 15 sediment samples from the Yellow estuary and showed that the concentration of PCDD/Fs was in 0.11-1.01 pg g^{-1}.

In Southeastern China, Zhang et al. (2009) showed that PCDD/Fs content in sediments from the Pearl River Delta was about 1.24-2.63 pg g^{-1} WHO-TEQ. Higher concentrations of PCDD/Fs were found in the sediments from Guangzhou and Dongguan. Ren et al. (2009) studied the patterns and sources of PCDD/Fs and DL-PCBs in surface sediments from the East River and founded that the contamination levels of PCDD/Fs ranged from 2.1 to 9.8 with mean concentration of 4.5 pg WHO98-TEQ g^{-1} and DL-PCBs ranged from 0.042 to 0.45 with mean concentration of 0.19 pg WHO98-TEQg^{-1}, respectively. Terauchi, H., S collected marine surface sediments from Hong Kong bay and tested the concentration of PCDD/Fs and DL-PCBs in these samples (Terauchi et al. 2009). In most sampling sites, concentrations of PCDDs were the highest, followed by DL-PCBs, PCDFs, PBDFs, PBDDs, MoBPCDDs and MoBPCDFs in this order. The concentration of PCDD/Fs in Hong Kong was 2500–6300 pg g^{-1} dry wt. From TEQs perspective, the average TEQs calculated from six locations in Hong Kong was 9.0–18.8 pg-TEQ g^{-1} dry wt. Although PCDDs levels were high in Hong Kong, major part of which was OCDD with extremely low TEF values.

3.2 Soil

In Northeastern China, PCDD/Fs were analyzed by Zhang, H. J. et al. in top soils collected from 30 sites in Daliao River Basin. The concentrations of PCDD/F ranged from 0.31 to 53.05 ng TEQ kg^{-1} dw (mean value, 7.00 ng TEQ kg^{-1} dw) in soils and the levels of PCDD/Fs contamination in paddy soils were generally higher than those of upland soils (Zhang et al. 2008). The author believed that PCDD/F contamination in paddy soils should be simultaneously attributed to the polluted water irrigation and the organochlorine pesticide application.

In Northern China, Chen studied 42 surface soil samples collected around Beijing in northern China in September 2002and found that the background dioxin level of the samples was about 0.85 pg TEQ g^{-1} (Chen et al. 2003).

In Eastern China, Liu, JS et al. (2009) studied the distribution of PCDD/Fs and DL-PCBs in the soil in a typical area of eastern China. 21 soil samples were evaluated and the range of WHO-TEQ values for the PCDD/Fs and DL-PCBs in 17 soil samples as background was 0.017-5.04 pg g^{-1} (dry weight, dw), with a mean value 0.967(+/-1.361) pg g^{-1} and medium value 0.348 pg g^{-1}, which indicates that the levels of PCDD/Fs and DL-PCBs over the major part of this district were low. However, the WHO-TEQ values (6.52-16.7 pg g^{-1} dw) for PCDD/Fs and DL-PCBs in soil samples near a known contaminated disassembly industrial park were much higher than that of the background investigation. The study of Xu et al. (2009) showed that the agricultural soil PCDD/F concentrations in 2007 ranged from 73.6 to 377 ng kg^{-1} (0.60-6.38 ng I-TEQ kg^{-1}) in the vicinity of a municipal solid waste incinerator.

During 2006-2007, the overall soil PCDD/F levels increased significantly, i.e., 33% and 39% for total concentration and I-TEQ(median value), respectively. The results of a congener-specific factor analysis between soils and dioxin emission sources suggested that diffuse sources would be mainly from open burning of wastes, traffic and hot water boilers, which were supposed the major responsible for the accumulation of PCDD/Fs in soils.

In Southern China, soil samples from rice and cotton fields in Hubei Province in 1994 were collected for analysis of pollutants. The fields were suspected of being polluted by pesticides and contaminated irrigation water. The mean PCDD/F concentrations in the samples were about 26.21 pg g^{-1} (0.11 pg TEQ g^{-1}) for samples from the cotton field and 95.55 pg g^{-1} (1.30 pg WHO-TEQ g^{-1}) for samples from the rice field (Wu et al. 1998, 2002). The congener profiles of the samples showed high proportions of OCDD, OCDF, and 1234678-HpCDD. The study of Zhao et al. (2006) determined the PCDD/Fs level in soil and associated biota samples collected from a polluted and wasted farmland in southern China. The pattern of PCDD/Fs and PCBs in soils and plants could reflect the original pollution source after transportation and biodegradation for 14 years and the consumption foods in this region such as foraging chicken eggs the total toxicity equivalent (TEQ) was up to 784 pg WHO-TEQ g^{-1} on basis lipid in foraging chicken eggs.

In Southeastern China, Zhang SK et al. (2009) showed that PCDD/PCDF concentration 17 soil samples are higher than 4 ng kg^{-1} but lower than 20 ng kg^{-1}, whereas the WHO1998-TEQ values calculated for the remaining 45 samples are lower than 4 ng kg^{-1} from the Pearl River Delta. This value is comparable to that for lightly polluted urban and industrial soil of other countries [e.g., 0.7–1.5 pg TEQ g^{-1} in Italy (Caserini et al. 2004), 1.1–11 pg TEQ g^{-1} in Switzerland (Schmid et al. 2005), and 0.13~24.20 pg TEQ g^{-1} in Spain (Schuhmacher et al. 2002)lower than the background level for soil in Tokyo (42.8 pg TEQ g^{-1}), and higher than the background level for unpolluted areas [e.g., 0.16 pg TEQ g^{-1} in Spain and 0.4–4.27 pg I-TEQ kg^{-1} in Japan (Sakurai et al. 2000)].

3.3 Ambient Air

There is still no regular dioxin-monitoring networks for ambient air in mainland China. Dioxin monitoring data for ambient air samples were only available for several cities in the published data. PCDD/Fs were monitored in the ambient air of Beijing, China by Li et al. (2008), from February to December 2006 to evaluate their concentrations, congener profiles and gas-particle partitioning. The PCDD/F concentrations for three different districts ranged from 275 to 10,780 fg m^{-3}, with an average of 4,355 fg m^{-3}. The I-TEQs value were 18-644 fg I-TEQ m^{-3}, with an average of 268 fg I-TEQ m^{-3}. Li et al. (2008) also measured the concentrations of PCDD/Fs and PBDD/Fs congeners in the ambient air of four districts in Shanghai. The mean atmospheric concentrations (TEQs) of total 2,3,7,8-PCDD/Fs and 2,3,7,8-PBDD/Fs were 8,031 fg m^{-3} (497.1 fg I-TEQ m^{-3}) and 1,358 fg m^{-3} (304.1 fg I-TEQ m-3) for Jiading District, 5,308 fg m^{-3} (289.0 fg I-TEQ m^{-3}) and 709 fg m^{-3} (146.9 fg I-TEQ m^{-3}) for Zhabei District, 4,014 fg m^{-3} (144.4 fg I-TEQ m^{-3}) and 1,239 fg m^{-3} (256.9 fg I-TEQ m^{-3}) for Pudong District, 3,348 fg m^{-3} (143.2 fg I-TEQ m^{-3}) and 699 fg m^{-3} (148.4 fg I-TEQ m^{-3}) for Huangpu District, respectively. Gas/particle partitioning of PCDD/Fs in ambient air was investigated in a satellite town in Eastern China by Xu et al. (2009) from April 2007 to January 2008. The results showed that the PCDD/Fs level in winter time was much higher than that in summer time and the value lied in the range of 86-3,030 fg m^{-3}. The lowest and highest level appeared in July 2007 and Jan 2008 respectively, which was comparable or slightly higher than other urban locations around the world.

PCDD/F and dioxin-like PCB were measured in 142 air samples of Hong Kong by Choi et al. 2008. The annual average PCDD/F and dioxin-like PCBs concentrations obtained for Hong Kong air at Tap Mun from January 2004 to March 2005 were (PCDD/F: 1,724; DL-PCB: 1,572 fg m^{-3}), Yuen Long (PCDD/Fs: 2,927; DL-PCB: 4,331 $\pm$ 1,962 fg m^{-3}) and Tsuen Wan (PCDD/F: 1,875; DL-PCB: 2,972 fg m^{-3}). The ambient air of Dalian, which is a tourism city in northeastern China, had a PCDD/F level of 2.72 fg TEQ m^{-3}, a value that is much lower than that for Beijing, Shanghai and Hong Kong. Most of the available data shown that the ambient air dioxin levels in Chinese cities were comparable to the values for the urban air of Spain (10–357 fg TEQ m^{-3}), Brazil (47–751 pg TEQ m^{-3}) and Athens (166.6–701.5 fg m^{-3}) (Abad et al. 2002, Assuncao et al. 2005, Moon et al. 2005.

4. Major Emission Source Analysis

4.1 Pentachlorophenol and Sodium Pentachlorophenate Products

High concentrations of OCDD, OCDF, and 1,2,3,4,6,7,8-HpCDD were found in PCP-Na products, ranging from 30 to 92 ng I-TEQ g -1 (Schecter et al., 1994; Bao et al., 1995). This range approached PCDD/F concentrations in Japanese PCP-Na produced by the hexachlorobenzene method (Seike et al., 2003). As for the congener pattern OCDD was the predominant congener, (OCDD > 1,2,3,4,6,7,8-HpCDD > OCDF > 1,2,3,4,6,7,8-HpCDD). This patter differed from that of the PCP-Na produced in Japan.

Because Na-PCP has been widely produced and used principally to control snailborne schistosomiasis in China, residual dioxins in the chemical have been released into the environment and have contributed notably to human exposure in areas where Na-PCP was used (Schecter et al., 1996; Hong et al., 2005). High PCDD/Fs concentrations were detected in sediment samples from Dongting Lake (Zheng et al., 1997b) and the Pearl River, where PCP-Na was sprayed (Zheng et al., 2001b). The concentrations of PCDD/Fs ranged from 128.3 to 326.9 pg I-TEQ g -1 (shown in fig.2). The study of Pan, J et al showed that impact of PCP-Na could last more than 10 years; the congener profiles of PCDD/Fs in the sediments were consistent with the profiles of main dominant PCDD/Fs in pentachlorophenol and sodium pentachlorophenate products in China. Wang, B's study for PCDD/Fs and dioxin-like compounds in sediments from the Haihe River also showed the dioxin pollution might be associated with industrial activities of pentachlorophenol (PCP) in the adjacent area.

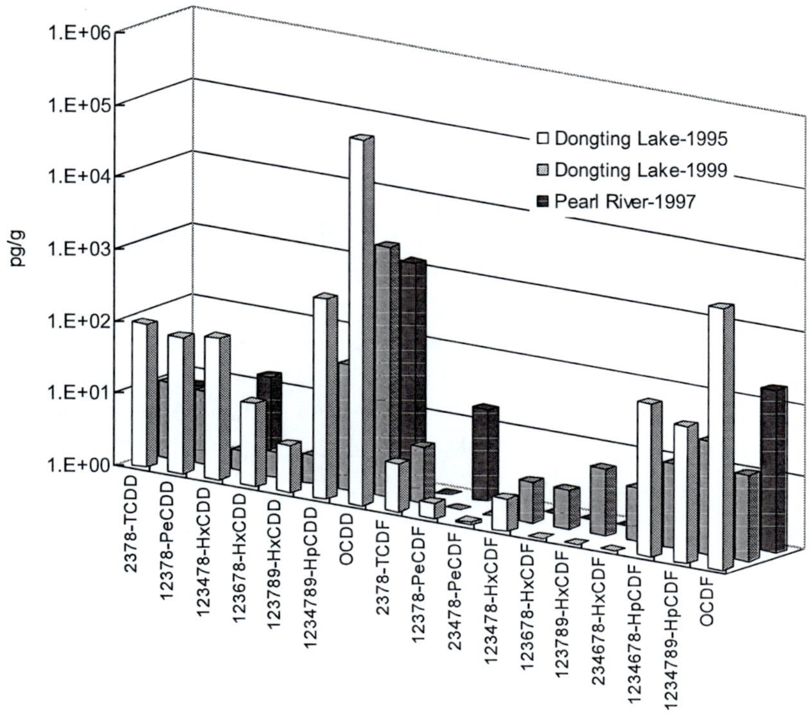

Figure 2. PCDD/F congener concentrations detected in Dongting Lake and Pearl River sediment samples

4.2 Waste Incineration

Incineration technologies for flammable hazardous and medical wastes have high priority in the National Program for Hazardous and Medical Waste Treatment Facility Construction Plan("the Program"). However, China still lacks the capacity to safely dispose of its hazardous and medical wastes, and as a result large amounts of hazardous waste have been stored temporarily. Only Shenzhen, Tianjin, Shenyang, and Hangzhou have centralized hazardous waste treatment facilities. The volume of hazardous waste has increased rapidly as the average annual growth rate was greater than 30% from 1999 to 2003. About 3.7×10^6 t of hazardous waste was disposed of in 2003.

Gao HC et al. (2009) studied the stack gas emissions of PCDD/Fs from 14 domestic-made hospital waste incinerators in China, whose burning capacities ranged from 5 to 25 ton d^{-1}. The results showed that the stack gas emissions of PCDD/Fs from HWIs exhibited a large variation (0.08-31.60 ng I-TEQ Nm^{-3}). Nine incinerators had the emission levels below the current emission standard in China (0.5 ng I-TEQ Nm^{-3}), while only two facilities exhibited emission levels below the European Union directive emission limit (0.1 ng I-TEQ Nm^{-3}). Chen T et al. (2009) studied the PCBs emission from a medical waste incinerator in China and they found that the total PCBs concentration of all homologues (mono- to deca-chlorinated homologues) in the flue gas ranged from 138.01 to 855.35 ng Nm^{-3} and the

WHO-TEQ value varied from 0.046 to 0.549 ng Nm^{-3} under the different operating conditions. It was estimated that 4.87 g I-TEQ of PCDD/Fs was annually released from hospital waste incineation to the atmosphere in China in 2006, which is much less the number of 427.4 g I-TEQ estimated in the preliminary official inventory.

Ni YW et al. (2009) studied the Emissions of PCDD/Fs from municipal solid waste incinerators in China and 19 commercial municipal solid waste (MSW) incinerators in China are investigated. The emission concentrations of PCDD/Fs were 0.042-2.461 ng TEQ Nm^{-3} with an average value of 0.423 ng TEQ Nm^{-3}. The emissions of PCDD/Fs from 16 MSW incinerators were below the MEP regulation level (1.0 ng I-TEQ Nm^{-3}, while only six systems have the dioxin emission levels below the limit established by the European Union Directive of 0.1 ng I-TEQ Nm^{-3}. The emission factors of PCDD/Fs from 19 MSW incinerators were calculated to be 0.169-10.72 ìg TEQ ton^{-1} MSW with an average value of 1.728 ìg I-TEQ ton^{-1} MSW. And The total amount of PCDD/Fs emitted from MSW incinerators to the atmosphere in China was estimated to 19.64 g TEQ $year^{-1}$ in 2006, which is much less the number of 125.8 g I-TEQ estimated in the preliminary official inventory.

4.3 Metallurgy

Ferrous and non-ferrous metal production China is one of the world's largest producers of ferrous and non-ferrous metals. Based on the large quantities of metals produced in 2002 in combination with emission factors given in the UNEP PCDD/Fs Toolkit (UNEP Chemicals, 2005), the potential PCDD/F emissions from metal production will be a very big number. Especially for the secondary metallurgy process, high strength of PCDD/Fs might be released out. The sediments in Southeastern China were also characterized by the elevated levels of PCDDs, especially OCDD. Source analysis revealed that PCDD/Fs in the sediments from Guangzhou were mainly from the secondary copper smelters and steel-making plants.

Ba T et al. (2009) investigated the emission factors and total emissions amounts of PCDD/Fs and DL-PCBs released from secondary aluminum and copper metallurgy industries in China. The results showed that the TEQ emission factor of PCDD/Fs is higher for secondary copper production, at 14,802 ng TEQ t^{-1} than for secondary aluminum production, at 2,650 ng TEQ t^{-1}. However, the TEQ emission factor of DL-PCBs of secondary aluminum production, with 193 ng TEQ t^{-1}, is higher than that of secondary copper production with 98.1 ng TEQ t^{-1} and the emission factors and the total emission amounts of the secondary aluminum and copper metallurgies in China stay in the middle level compared to values reported for other countries. The total estimated emissions of PCDD/Fs released to air from the production of 2.75 million tons secondary aluminum and 2 million tons secondary copper in 2007 are 7.3 and 37.5 g TEQ yr^{-1}, respectively. The corresponding DL-PCBs total emission amounts being 0.53 and 0.2 g TEQ yr^{-1} respectively. The estimated dioxin emission strength was much less than the values calculated in the inventory (133.5 g for aluminum and 403g for copper in I-TEQ).

As for the secondary zinc and lead production studied by Ba T et al. (2009) showed the toxic equivalent quantity (TEQ) emission factor of PCDD/Fs and DL-PCBs released into the environment is higher for secondary zinc production, at 52 298.02 ng TEQ ton^{-1} than for secondary lead production, at 646.05 ng TEQ ton^{-1}. Based on the emission factor and production level, the total estimated emission amounts of PCDD/Fs and DL-PCBs in both stack gas and fly ash released into the environment from secondary zinc and lead production is estimated to be at least 2.76 and 0.42 g TEQ yr^{-1}, respectively. The DL-PCBs contribute 2.8% and 0.6% of the total emission from secondary zinc and lead plants, respectively. The estimated dioxin emission strength was just in the same fold as the values calculated in the inventory (8 g for aluminum and 13.4 g for copper in I-TEQ).

4.4 E-waste

Pollution from illegal e-waste recycling is a special problem in China and other developing Asian countries. High concentrations of PCDD/Fs were detected in e-waste samples collected near the open burning sites in the town of Guiyu, Guangdong Province (Fig.3) (Luksemburg et al., 2002). PCDD/F concentrations in samples collected from the blackened e-waste and burning ash were as high as 469 and 5678 pg I-TEQ g–1, respectively. PCDD/F concentrations in river sediments collected near this e-waste recycling site were 21~39000 pg I-TEQ g–1. The congener patterns of these sediments were quite different from those of the Yangtze River, which had very high HpCDFs, PeCDFs, TCDD, and TCDF, and relatively lower OCDD, similar to the e-waste and ash. Therefore, the PCDD/Fs detected in the local river sediments may have been derived from the illegal e-waste treatment.

Shen et al. (2009) characterized the dioxin-like compounds in Taizhou area, one of the largest e-waste recycling centers in China, using both chemical analysis and in vitro bioassay. Raw soil extracts from all locations induced significant AhR activities, where the TEQ value ranged from 5.3 to 210 pg g^{-1} dry weight soil (pg g^{-1} dw). The total concentrations of 17 PCDD/Fs, 36 PCBs and 16 PAHs varied from 210 to 850 pg g^{-1} dw, 11 to 100 ng g^{-1} dw, and 330 to 20,000 ng g^{-1} dw, respectively. Profile characterization of the target analytes revealed that there were similar sources originating from the crude dismantling of electric power equipments and the open burning of e-waste. Liu et al. (2008) analyzed the PCDD/Fs in 20 soil samples collected from an electronic waste recycling site and its vicinage towns. The toxic equivalency (TEQ) of PCBs and PCDD/Fs detected in E-waste recycling site is significant higher than those in the vicinage samples was up to 789 ng g^{-1} dry weight (dw).

The study of Li, YM. et al. (2008) indicated that high level of PCDD/Fs, PCBs and PBEDs could be also detected in the ambient air of Taizhou, an E-waste dismantling area of southeast China. The I-TEQs for PCDD/Fs were in the range of 0.20-3.45 pg m^{-3}, with an average of 1.10 pg m^{-3}. The Sigma PCBs concentrations and TEQs ranged from 4.23 to 11.35 ng m^{-3}, 0.050 to 0.859 pg TEQ m^{-3}, respectively. The concentrations of Sigma PBDEs ranged from 92 to 3086 pgm^{-3}, with an average of 894 pg m^{-3}. The pollution levels of PCDD/Fs, PCBs and PBDEs were higher than other urban sites, which may be associated with the E-waste dismantling activities. All of the results discussed above suggest that e-waste recycling may be an important source of PCDD/Fs in some areas of Guangdong and Zhejiang provinces, where large amounts of e-waste were illegally recycled recently.

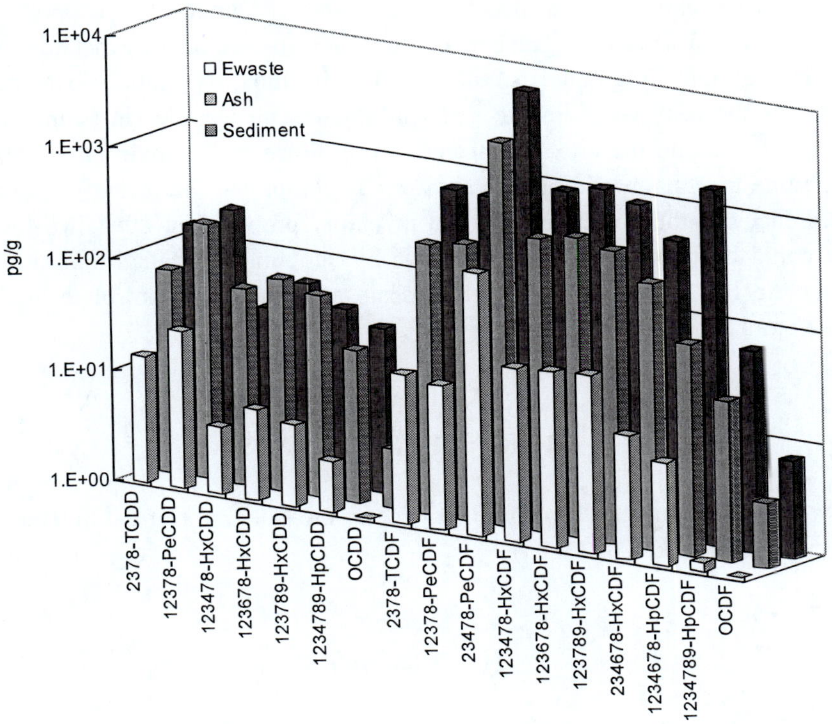

Figure 3. PCDD/F congener concentrations in e-waste, ash, and sediment samples collected in Guiyu, Guangdong province.

4.5 Open Burning

Zhang, QJ. et al. (2008) estimated the PCDD/Fs emissions from open burning of crop residues in China between 1997 and 2004 and their research results showed that annual emissions of PCDD/Fs from open burning of crop residues in each province of China mainland between 1997 and 2004 were estimated to be ranged from 1,380 to 1,520 g I-TEQ/yr, with the average of 1,500±80 g I-TEQ/yr, which contributed to approximately 10% similar to 20% of the total emissions in China, and should be concerned much in the later years.

CONCLUSIONS

The development of dioxin monitoring capacity and preliminary inventory of dioxin emission in China were introduced. The dioxin pollution level in the major environmental media and dioxin emission from PCP-Na production and usage, municipal, hazardous, and medical waste incineration; illegal e-waste recycling; and ferrous and non-ferrous metal production were reviewed for the published data from 1999 to 2010. It was shown that the dioxin monitoring capacity had been enhanced largely under the support of Chinese government in the past ten years, especially after the signature of the Stockholm Convention on Persistent Organic Pollutants in 2001. As for the dioxin pollution level in China the published data showed that sediment, soil and ambient air samples in China were slightly polluted by dioxins and the dioxin concentration in those major environmental media were just at the same level as that in other countries. By comparing the recently published dioxin monitoring data and the preliminary dioxin inventory proposed in 2007, we found that the inventory could be improved in some aspects by adopting the actual emission data when considering the big differences lied in the scale and pollution control technology in the industrial sectors concerned in China.

Acknowledgments

This work was financially supported, in part, by the National Natural Science Foundation of China (50708110) and (20977105).

References

Abad, E., Caixach, J., Rivera, J., Gustems, L., Massague, G., Puig, O., 2002. Surveillance programme on dioxin levels in ambient air sites in Catalonia (Spain). *Chemosphere* 49, 697–702.

Assuncao, J.V., Pesquero, U.R., Bruns, R.E., Carvalho, L., 2005. Dioxins and furans in the atmosphere of Sao Paulo City, Brazil. *Chemosphere* 58, 1391–1398.

Ba, T., M. H. Zheng, et al. 2009. Estimation and characterization of PCDD/Fs and dioxin-like PCB emission from secondary zinc and lead metallurgies in China. *Journal of Environmental Monitoring* 11, 867-872.

Bao, Z.C.,Wang, K., Kang, J.X., Zhao, L.W., 1995. Analysis of polychlorinated dibenzo-p-dioxins and polychlorinated dibenzofurans in pentachlorophenol and sodium pentachlorophenate. *Environ. Chem.* 14, 317–321 (in Chinese).

Caserini, S., Cernuschi, S., Giugliano, M., Grosso, M., Lonati, G., Mattaini, P., 2004. Air and soil dioxin levels at three sites in Italy in proximity to MSW incineration plants. *Chemosphere* 54, 1279–1287.

Chen, J. W., H. M. Zhao, et al. 2006. Atmospheric PCDD/F and PCB levels implicated by pine (Cedrus deodara) needles at Dalian, China. *Environmental Pollution* 144(2): 510-515.

Chen, T., X. D. Li, et al. 2009. Polychlorinated biphenyls emission from a medical waste incinerator in China. *Journal of Hazardous Materials* 172(2-3): 1339-1343.

Chen, Z.S., Li, W., Li, C.Q., Huang, P., Liu, G.Y., Zhou, Z., 2003. Background levels of PCDD/Fs in soil of Beijing area, China. *Organohal. Compd.* 62, 495–498.

Choi, M. P. K., S. K. M. Ho, et al. 2008. PCDD/F and dioxin-like PCB in Hong Kong air in relation to their regional transport in the Pearl River Delta region. *Chemosphere* 71, 211-218.

Fattore, E., Vigano, L., Mariani, G., Guzzi, A., Benfenati, E., Fanelli, R., 2002. Polychlorinated dibenzo-p-dioxins and dibenzofurans in River Po sediments. *Chemosphere* 49, 749–754.

Frignani, M., Bellucci, L.G., Carraro, C., Raccanelli, S., 2001. Polychlorinated biphenyls in sediments of the Venice Lagoon. *Chemosphere* 43, 567–575.

Gao, H. C., Y. W. Ni, et al. 2009. Stack gas emissions of PCDD/Fs from hospital waste incinerators in China. *Chemosphere* 77, 634-639.

Hong, H.C., Zhou, H.Y., Luan, T.G., Lan, C.Y., 2005. Residue of pentachlorophenol in freshwater sediments and human breast milk collected from the Pearl River Delta, *China. Environ. Int.* 31, 643–649.

Hosomi,M.,Matsuo, T., Dobashi, S., Katou, S., Abe, H., 2003. Survey of dioxins in Tokyo Bay bottom sediment. Mar. *Pollut. Bull.* 47, 68–73.

Hu, J.Y., Wan, Y., Shao, B., Jin, X., An, W., Jin, F., Yang, M., Wang, X.J., Sugisaki, M., 2005. Occurrence of trace organic contaminants in Bohai Bay and its adjacent Nanpaiwu River, North China. Mar.*Chem.* 95, 1–13

Hui, Y. M., M. H. Zheng, et al. 2009. PCDD/Fs and DL-PCBs in Sediments from Yellow Estuary and Yangtze Estuary, China. *Bulletin of Environmental Contamination and Toxicology* 83, 614-619.

Jin J , Hao P , Tang XY. 2004. An inventory of potential PCDD and PCDF emission sources in the mainland of China [J]. *Organohalogen Compound*, 66:852-858

Li, H. R., H. L. Feng, et al. 2008. The PCDD/F and PBDD/F pollution in the ambient atmosphere of Shanghai, China. *Chemosphere* 70, 576-583.

Li, K., H. W. Yin, et al. 2007. Polychlorinated dibenzo-p-dioxins, dibenzofurans and dioxinlike biphenyls in sediments from the Suzhou Creek, China. *Bulletin of Environmental Contamination and Toxicology* 79, 432-436.

Li, Y. M., G. B. Jiang, et al. 2008. Concentrations, profiles and gas-particle partitioning of PCDD/Fs, PCBs and PBDEs in the ambient air of an E-waste dismantling area, southeast China. *Chinese Science Bulletin* 53, 521-528.

Li, Y. M., G. B. Jiang, et al. 2008. Concentrations, profiles and gas-particle partitioning of polychlorinated dibenzo-p-dioxins and dibenzofurans in the ambient air of Beijing, China. *Atmospheric Environment* 42, 2037-2047.

Liu, H. X., Q. F. Zhou, et al. 2008. E-waste recycling induced polybrominated diphenyl ethers, polychlorinated biphenyls, polychlorinated dibenzo-p-dioxins and dibenzo-furans pollution in the ambient environment. *Environment International* 34, 67-72.

Liu, H. X., Q. H. Zhang, et al. 2007. Occurrence of polychlorinated dibenzo-p-dioxins, dibenzofurans and biphenyls pollution in sediments from the Haihe River and Dagu Drainage River in Tianjin city, China. *Chemosphere* 68, 1772-1778.

Liu, H. X., Q. H. Zhang, et al. 2007. Occurrence of polychlorinated dibenzo-p-dioxins, dibenzofurans and biphenyls pollution in sediments from the Haihe River and Dagu Drainage River in Tianjin city, China. *Chemosphere* 68, 1772-1778.

Luksemburg, W.J., Mitzel, R.S., Peterson, R.G., Hedin, J.M., Maier, M., Schuld, M., Zhou, H., Wong, A.S., 2002. Polychlorinated dibenzodioxins and dibenzofurans (PCDDs/PCDFs) levels in environmental and human hair - samples around an electronic waste processing site in Guiyu, Guangdong Province, China. *Organohalogen Compd.* 55, 347–349.

Luksemburg, W.J., Mitzel, R.S., Peterson, R.G., Hedin, J.M., Maier, M.,Schuld, M., Zhou, H., Wong, A.S., 2002. Polychlorinated dibenzodi-oxins and dibenzofurans (PCDDs/PCDFs) levels in environmental and human hair samples around an electronic waste processing site in Guiyu, Guangdong Province, China. *Organohal. Compd.* 55, 347–349.

Luksemburg, W.J., Mitzel, R.S., Zhou, H., Hedin, J.M., Silverbush, B.B., Wong, A.S., 1996. Polychlorinated dioxins and dibenzofurans in environmental samples from China. *Organohal. Compd.* 28, 262–266.

Luo, J. P., M. Ma, et al. 2009. Characterization of aryl hydrocarbon receptor agonists in sediments of Wenyu River, Beijing, China. *Water Research* 43, 2441-2448.

Martens, D., Zhang, A., Jiang, X., Chen, J., Gawlik, B.M., Henkelmann, B., Schramm, K.W., Kettrup, A., 2000. Polychlorinated dibenzo-p-dioxins and dibenzofurans, pentachlorophenol, pentachloroanisole and hexachlorobenzene in sediments of the Yangtse River and the Liao-He River in China. *Organohal. Compd.* 46, 431–434.

Marvin, C., Alaee, M., Painter, S., Charlton, M., Kauss, P., Kolic, T., MacPherson, K., Takeuchi, D., Reiner, E., 2002. Persistent organic pollutants in Detroit River suspended sediments, polychlorinated dibenzo-p-dioxins and dibenzofurans, dioxin-like polychlorinated biphenyls and polychlorinated naphthalenes. *Chemosphere* 49, 111–120.

Moon, H.B., Lee, S.J., Choi, H.G., Ok, G., 2005. Atmospheric deposition of polychlorinated dibenzo-p-dioxins (PCDDs) and dibenzofurans (PCDFs) in urban and suburban areas of Korea. *Chemosphere* 58, 1525–1534.

Muller, J.F., Gaus, C., Prange, J.A., Papke, O., Poon, K.F., Lam, M.H., Lam, P.K., 2002. Polychlorinated dibenzo-p-dioxins and polychlorinated dibenzofurans in sediments from Hong Kong. Mar. *Pollut. Bull.* 45, 372–378.

National Implementation Plan (NIP) for the Stockholm Convention on Persistent Organic Pollutants of the Peoples Republic of China, April 2007, *http://www.pops.int.*

Ni, Y. W., H. J. Zhang, et al. 2009. Emissions of PCDD/Fs from municipal solid waste incinerators in China. *Chemosphere* 75, 1153-1158.

Onodera, S., Sugimoto,M., Takagi, T., Tanaka, K., 1999. Polychlorinated dibenzo-p-dioxins and dibenzofurans, their concentrations and profiles in sediments in the Tama River, Japan. Jpn. *J. Toxicol. Environ. Health* 45, 1–7.

Pan, J., Y. L. Yang, et al. 2010. Polychlorinated biphenyls, polychlorinated dibenzo-p-dioxins and dibenzofurans in marine and lacustrine sediments from the Shandong Peninsula, China. *Journal of Hazardous Materials* 176, 274-279.

Qiao, M., Y. Y. Chen, et al. 2006. Identifi,cation of Ah receptor agonists in sediment of Meiliang Bay, Taihu Lake, China. *Environmental Science & Technology* 40, 1415-1419.

Ren, M., P. A. Peng, et al. 2009. Patterns and sources of PCDD/Fs and DL-PCBs in surface sediments from the East River, China. *Journal of Hazardous Materials* 170, 473-478.

Sakurai, T., Kim, J.G., Suzuki, N., Matsuo, T., Li, D.Q., Yao, Y.A., Masunaga, S., Nakanishi, J., 2000. Polychlorinated dibenzo-p-dioxins and dibenzofurans in sediment, soil, fish, shellfish and crab samples from Tokyo Bay area, Japan. *Chemosphere* 40, 627–640.

Schecter, A., Furst, P., Furst, C., Papke, O., Ball, M., Ryan, J.J., Hoang, D.C., Le, C.D., Hoang, T.Q., Cuong, H.Q., Phuong, N.T.N., Phiet, P.H., Beim, A., Constable, J., Startin, J., Samedi, M., Seng, Y.K., 1994. Chlorinated dioxins and dibenzofurans in human tissue from general populations - a selective review. *Environ. Health Persp.* 102 SU (Suppl. 1), 159–171.

Schecter, A.J., Li, L.J., Ke, J., Furst, P., Furst, C., Papke, O., 1996. Pesticide application and increased dioxin body burden in male and female agricultural workers in China. *J. Occup. Environ. Med.* 38, 906–911.

Schmid, P., Gujer, E., Zennegg, M., Bucheli, T.D., Desaules, A., 2005. Correlation of PCDD/F and PCB concentrations in soil samples from the Swiss soil monitoring network (NABO) to specific parameters of the observation sites. *Chemosphere* 58, 227–234.

Schuhmacher, M., Agramunt, M.C., Rodriguez-Larena, M.C., Diaz-Ferrero, J., Domingo, J.L., 2002. Baseline levels of PCDD/Fs in soil and herbage samples collected in the vicinity of a new hazardous waste incinerator in Catalonia, Spain. *Chemosphere* 46, 1343–1350.

Schuhmacher, M., Agramunt, M.C., Rodriguez-Larena, M.C., Diaz-Ferrero, J., Domingo, J.L., 2002. Baseline levels of PCDD/Fs in soil and herbage samples collected in the vicinity of a new hazardous waste incinerator in Catalonia, Spain. *Chemosphere* 46, 1343–1350.

Seike, N., Takashi, O.,Masako, U., Takumi, T., Nobuyuki, T., 2003. Temporal change of polychlorinated dibenzo-p-dioxin, dibenzofuran and dioxin-like polychlorinated biphenyl sources in paddy soils. *J. Environ. Chem.* 13, 117–131 (in Japanese).

Shen, C. F., Y. X. Chen, et al. 2009. Dioxin-like compounds in agricultural soils near e-waste recycling sites from Taizhou area, China: Chemical and bioanalytical characterization. *Environment International* 35, 50-55.

Takigami, H., Sakai, S., Brouwer, A., 2005. Bio/chemical analysis of dioxin-like compounds in sediment samples from Osaka Bay, Japan. *Environ. Technol.* 26, 459–469.

Terauchi, H., S. Takahashi, et al. 2009. Polybrominated, polychlorinated and monobromo-polychlorinated dibenzo-p-dioxins/dibenzofurans and dioxin-like polychlorinated biphenyls in marine surface sediments from Hong Kong and Korea. Environmental Pollution 157, 724-730.

Tian HH, Hai Y, Yue RS. 2002. *Twelve persistent organic pollutants (POPs) in China.* Workshop on Environmental Monitoring of Persistent Organic Pollutants in East Asian Countries, Tokyo: 40-47

UNEP Chemicals, 2005. *Standardized Toolkit for Identification and Quantification of Dioxin and Furan Releases*. UNEP Chemicals, Geneva

Wang, B., G. Yu, et al. 2009. CALUX Bioassay of Dioxin-Like Compounds in Sediments from the Haihe River, China. *Soil & Sediment Contamination* 18, 397-411.

Wen, S., Y. Hui, et al. 2008. Polychlorinated dibenzo-p-dioxins (PCDDs) and dibenzofurans (PCDFs) in surface sediment and bivalve from the Changjiang Estuary, China. Chinese *Journal of Oceanology and Limnology*. 26, 35-44.

Wu, W.Z., Schramm, K.W., Henkelmann, B., Xu, Y., Yediler, A.,Kettrup, A., 1997. PCDD/F-s, PCBs, HCHs and HCB in sediments and soils of Ya–Er Lake area in China, results on residual levels and correlation to the organic carbon and particle size. *Chemosphere* 34, 191–202.

Wu, W.Z., Schramm, K.W., Henkelmann, B., Xu, Y., Zhang, Y.Y., Kettrup, A., 1998. Survey on PCDDs and PCDFs in sediments and soils in Ya–Er Lake area, China. *Chin. J. Oceanol. Limnol.* 16, 45–53.

Wu, W.Z., Schramm, K.-W., Xu, Y., Kettrup, A., 2002. Contamination and distribution of polychlorinated dibenzo-p-dioxins and dibenzofurans (PCDD/F) in agriculture fields in Ya–Er Lake area, China. *Ecotoxicol. Environ. Safety* 53, 141–147.

Xu, M. X., J. H. Yan, et al. 2009. Agricultural soil monitoring of PCDD/Fs in the vicinity of a municipal solid waste incinerator in Eastern China: Temporal variations and possible sources. *Journal of Hazardous Materials* 166, 628-634.

Xu, M. X., J. H. Yan, et al. 2009. Gas/particle partitioning of atmospheric PCDD/Fs in a satellite town in Eastern China. *Chemosphere* 76, 1540-1549.

Zhang, H. J., Y. W. Ni, et al. 2008. Polychlorinated dibenzo-p-dioxins and dibenzofurans in soils and sediments from Daliao River Basin, China. *Chemosphere* 73, 1640-1648.

Zhang, P., J. M. Song, et al. 2009. Persistent organic pollutant residues in the sediments and mollusks from the Bohai Sea coastal areas, North China: An overview. *Environment International* 35, 632-646.

Zhang, Q. H. and G. B. Jiang 2005. Polychlorinated dibenzo-p-dioxins/furans and polychlorinated biphenyls in sediments and aquatic organisms from the Taihu Lake, China. *Chemosphere* 61, 314-322.

Zhang, Q., J. Huang, et al. 2008. Polychlorinated dibenzo-p-dioxins and dibenzofurans emissions from open burning of crop residues in China between 1997 and 2004. *Environmental Pollution* 151, 39-46.

Zhang, S. K., P. A. Peng, et al. 2009. PCDD/PCDF pollution in soils and sediments from the Pearl River Delta of China. *Chemosphere* 75, 1186-1195.

Zhao, X. R., M. H. Zheng, et al. 2006. Evidence for the transfer of polychlorinated biphenyls, polychlorinated dibenzo-p-dioxins, and polychlorinated dibenzofurans from soil into biota. *Science of The Total Environment* 368, 744-752.

Zheng MH, Sun YZ, Liu WB, 2008, *Dioxin Emission Inventory in China,* Environmental Science Press of China, Beijing: 9-14

Zheng, M., Zhang, B., Bao, Z.C., Yang, H., 2003. Analysis of PCDD/Fs from sediments of Dongting Lake in China. *Organohal. Compd.* 62, 190–192.

Zheng, M.H., Bao, Z.C., Wang, K.O., Yang, H., Xu, X.B., 1997.Polychlorinated dibenzo-p-dioxins and dibenzofurans in lake sediments from Chinese schistosomiasis areas. *Bull. Environ. Contam.Toxicol.* 59, 653–656.

Zheng, M.H., Bao, Z.C., Wang, K.O., Yang, H., Xu, X.B., 1997b. Polychlorinated dibenzo-p-dioxins and dibenzofurans in lake sediments from Chinese schistosomiasis areas. *Bull. Environ. Contam. Toxicol.* 59, 653–656.

Zheng, M.H., Chu, S.G., Sheng, G.Y., Min, Y.S., Bao, Z.C., Xu, X.B., 2001. Polychlorinated dibenzo-p-dioxins and dibenzofurans in surface sediments from Pearl River Delta in China. *Bull. Environ. Contam. Toxicol.* 66, 504–507

Zheng, M.H., Chu, S.G., Sheng, G.Y., Min, Y.S., Bao, Z.C., Xu, X.B., 2001b. Polychlorinated dibenzo-p-dioxins and dibenzofurans in surface sediments from Pearl River Delta in China. *Bull. Environ. Contam. Toxicol.* 66, 504–507

In: Pollution in China
Editor: Michael I. Chang

ISBN: 978-1-61122-022-3
©2011 Nova Science Publishers, Inc.

Chapter 4

AIR POLLUTION FROM TRANSPORT SECTOR IN CHINA AND POLICIES TOWARD A SUSTAINABLE FUTURE

Ji Han*

Integrated Research System for Sustainability Science, The University of Tokyo, Tokyo, Japan

ABSTRACT

Transportation is a leading sector for energy consumption together with associated air pollutant and greenhouse gas (GHG) emissions, and one of the most difficult sources to control. In the worldwide scope, such issues as how to reduce the consumption of non-renewable energy resources, what kind of effective remedies can be taken to mitigate air pollution and GHG emissions from transport sector have been paid more and more attentions not only by researchers but also by policy makers. Asian developing country like China, with the expected increase in levels of motorization and further economic growth, would eventually have to target air pollution control and low carbon transport more vigorously than before in the short as well as the long term.

To grasp the status of transport-related energy and environmental problems and their future trends in China so that proper policies could be made to achieve a sustainable development, in this chapter firstly the contribution of transport to air pollution in a local scale together with global warming among all the industrial sectors is investigated. Secondly, the inventories of air pollutant emissions are evaluated from inter-city transport including four modes such as railway, road, waterway and airway, and from urban transport consisting of private car, bus, trolley, taxi and rail transit. Thirdly, the strategies and policies relevant to energy conservation and emission control are summarized. More importantly, a system dynamics model is developed for quantitative policy assessment and projection of air pollution mitigation potential up to 2030.

* Research Fellow, D.Eng., Integrated Research System for Sustainability Science, The University of Tokyo, 7-3-1 Hongo, Bunkyo-ku, Tokyo, Japan 113-8654, E-mail: hanji_cn@hotmail.com.

1. INTRODUCTION

Transportation is a leading sector for energy consumption together with associated air pollutants and greenhouse gas (GHG) emissions, and one of the most difficult sources to control. In 2004, among other sectors, transport accounted for 26% of world's total energy use, and 23% of total energy-related GHG emissions. While in developing countries, where globalization is expanding trade flows, and rising income level is amplifying demand for motorized mobility, transport energy use is increasing faster than that in developed nations, and projected to grow from 31% in 2002 to 43% of world total transport energy use by 2025 (International Panel on Climate Change, 2007). Accordingly, transport emissions are poised to soar with the development of economy and motorization. Nowadays, the studies on how to achieve a sustainable development in urbanization, motorization and environment have become one of the most important focuses in both developed and developing countries especially under the context of global warming. China, characterized by the world's largest population, is one of the most rapidly growing countries and the second largest energy consumer, who would exceed USA to become the largest in the near future. Its annual GDP growth rate is around 10% during the last two decades. And the total energy consumption in 2000 is about 1300 Mtce, equivalent to 1/5 of OECD and 1/10 of world total. Meanwhile, China is also one of the worst environmental polluters with the largest SOx emissions and the largest CO2 emissions in 2010. It is widely recognized that the urbanization and motorization in China are on a path neither socially nor environmentally sustainable. Therefore, because of its great potential impacts, the study on China's sustainable motorization and corresponding strategies and policies to reduce environmental load especially in transport sector is vital not only for China itself but also for the rest of the world.

In the international scope, a variety of studies have been conducted in transport sector especially in developed nations concerning urbanization, motorization, related energy consumption and air pollutant emissions together with corresponding policies and strategies. For examples, the World Conference on Transport Research Society and Institute for Transport Policy Studies (2004) investigated the history and trends of motor vehicle ownership by world regions and analyzed the local air pollution, global warming and noise issues derived from motor vehicles in various metropolises. World Business Council for Sustainable Development (2004) extrapolated the regionalized car stock trajectory till 2030 by using region-specific starting conditions and estimated the related emissions. International Energy Agency (2001) built scenarios to predict automobile ownership and its implications for oil market and CO2 emissions in three large Asian economies till 2020. Meyer at al. (2007) projected passenger car stock and associated CO2 emissions from 11 world regions by adopting a multi-model approach. Kuhns et al. (2004) assessed the air pollutant emission factors for on-road gasoline and diesel engine vehicles in Las Vegas of USA based on remote sensing techniques. In China's case, He (2005) employed a scenario analysis to estimate the road transport development and CO2 emissions inventories under different strategies. Wang et al. (2006) used the International Vehicle Emission (IVE) model to estimated vehicular pollutants emissions in Shanghai city. Cai and Xie (2007) used the Computer Program to Calculate Emissions from Road Transport (COPERT) III model to evaluate motor vehicle's emissions from 1980 to 2005. However, most of the studies treat the transport evolution in a country or region as a whole rather than a systematic mode-wise analysis, investigate transport policies relying on qualitative and theoretical abstraction rather than quantitative evaluation of their effect on energy conservation and pollutant reduction, and project future trend of development by elasticity-based methods rather than considering the interactions of each component in the complicated system. This kind of analysis is more doubtable for those developing countries with less historical data but rapid development. Taking China as an example, socio-economic inequalities exist widely between the eastern and western and between the urban and rural areas. Urban and eastern parts of China are more developed with relatively high income and motorization level, while the situation in rural and western areas is the other way round.

Against the background and concerning the issues of mitigation of air pollutant and CO2 emissions especially from the transportation industry in China, we stare at the major transport modes that are expected to have significant impacts on local and global environment, such as inter-city passenger and freight transport, private car stock and urban public transport, etc. Through developing quantitative analysis models including system dynamics model, regression model and linear expenditure model, etc., we evaluate CO2 mitigation potential, atmospheric pollution, notably CH4, CO, Non-Methane Volatile Organic Compounds (NMVOC), NOx and SO2, and appropriate policy implications for future sustainability.

In order to fulfill the tasks stated above, studies of each section are carried out and organized in the structure showed in Figure 1. First, section 1 investigates the mechanism and process by which socio-economic development, urbanization and motorization come to act on environment in several aspects in particular focuses on energy use and air pollutant emissions. And then, the contributions of transportation to total energy consumption among all the industries during the period of 1995-2007 are measured. Next, section 2 analyzes the historical change of both inter-city and urban transport by taking into account the characteristics of major modes, and investigates the dominant factors that influence modal share and consequent transport-related energy consumption and pollutant emissions. In section 3, from the view point of policy, the currently implemented laws, policies and regulations categorized in three levels that aim at energy conservation and environmental protection are summarized. Finally, a system dynamics model and linear expenditure modal are proposed on the basis of historical data, plans and strategies for future projection of transport development and mitigation potential of related atmospheric pollution as well as CO2 emissions up to 2030. The model highlights the simulation of interactions among socio-economic driving forces such as regional economic inequality, population migration, policy influences, energy use, and air pollution with special attention to urban and rural differences in China, which are seldom considered in the previous studies.

1.1. Interactions of Urbanization, Motorization and Air Pollution

It is well known that environmental impacts and socioeconomic activities are closely interlinked. As illustrated in Figure 2, rapid economic growth would cause the average income level increase and disparity among regions and between urban and rural areas. Accordingly, a large amount of migrations would proceed in response to the difference in expected income (Harris and Todaro, 1970; Stiglitz, 1974). With more and more people move from rural to urban, the build-up land in cities must sprawl consequently which is usually

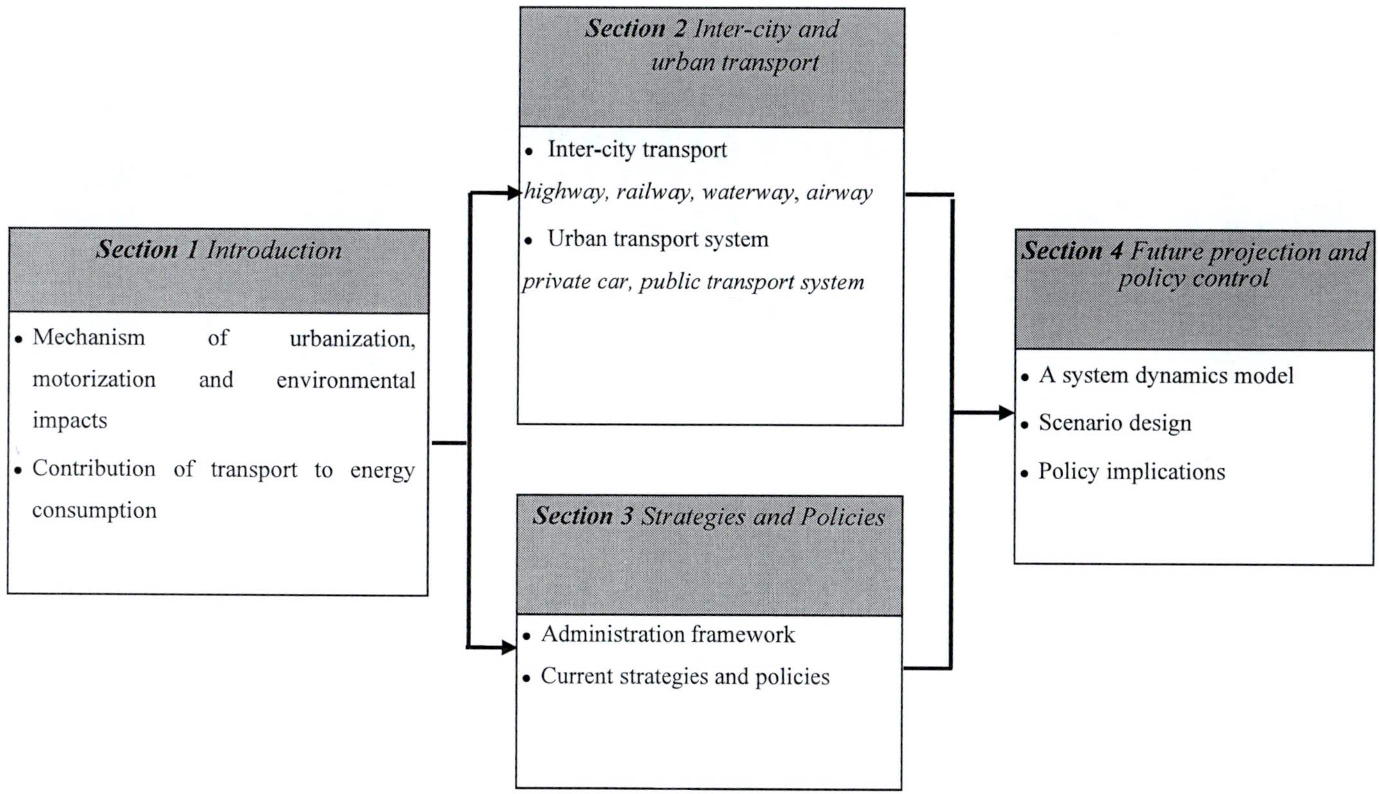

Figure 1. Section structure of this chapter.

accepted as the urbanization process. In China, it is widely known that since the economic reform in 1978, uneven regional development and political factors have created an environment for massive migration in China. From 1978 to 2000, the urbanization level in terms of the ratio of urban population to the nation's total grew from 18% to 36%, while the total urban population increased from about 170 million to 460 million people. During the period of 1995 to 2000, the total migration amounted to 128 million people. However, in developing countries urban growth is recognized as extensive rather than intensive, which results in the increasing demand for construction of infrastructures like roads, rails and bridges, etc. Insufficient financial investment and planning of urban land together with transport would cause longer trips and lower accessibility. In another hand, due to the increasing mobility demand from globalization as well as the rapid growth of income level, the use of private car and motorcycle is expected to soar especially in developing countries. For example, since the beginning of 21st century, the private car stock in China has increased by 2.9 times from 1.4 million in 2000 to 4.1 million in 2005, while the per capita GDP increased by only 1.2 times in the same period (National Bureau of Statistics, 2006; Department of Urban & Rural Society and Economic Statistics, 2006). It is reasonable to imagine that when the per capita GDP in China approaches the level of developed countries, the private car stock will reach a significant level. As a result, both the urbanization and motorization generate a number of environmental impacts such as land use change, increasing demand for domestic water resource, housing, energy use, and atmospheric pollution of local effect and CO_2 emissions of global impact. Among all the problems, this chapter focuses on the energy and air pollution from transport in China.

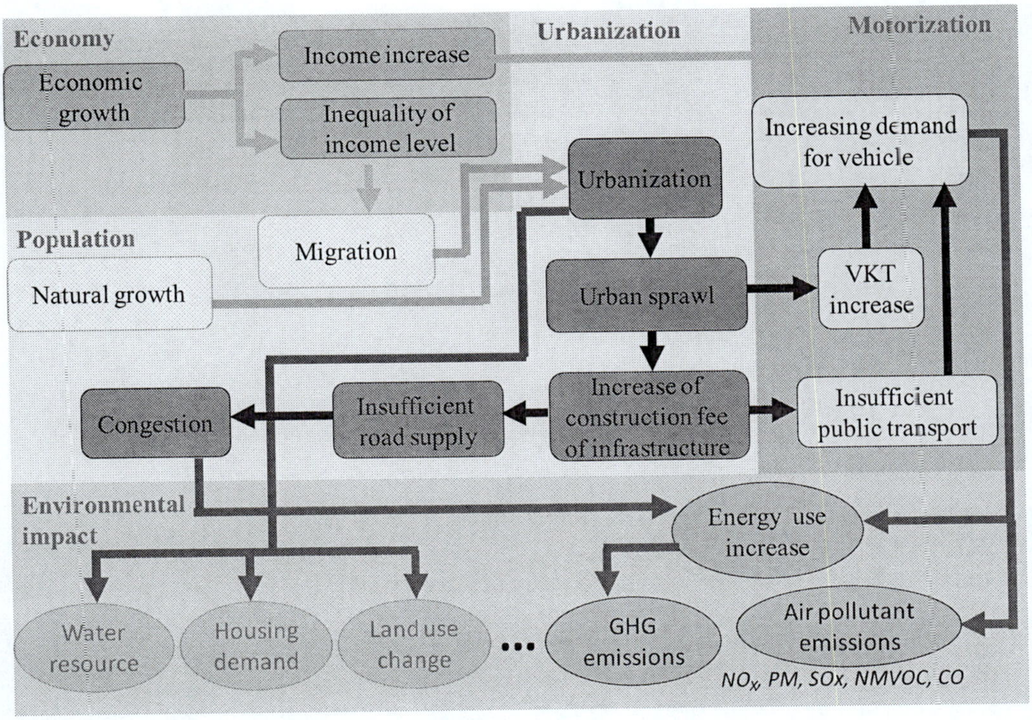

Figure 2. Mechanism of air pollution generated from transport.

1.2. Contribution of Transport to Energy Consumption

Energy consumption in the transport sector includes the energy consumed in moving people and goods by road, rail, air, water, and pipeline. Growth of economic activity and population are the key factors for transportation sector energy demand. Economic growth stimulates increases in industrial output, which requires the movement of raw materials to manufacturing sites, as well as the movement of manufactured goods to end users. Firstly, when looking at the position of China in the whole world's primary energy consumption, Figure 3 gives out an image of the top 10 largest energy consumers in 2007. China ranked the 2nd, just behind USA, with total energy use amounted to 2800 million tons of coal equivalent (MTCE) (Energy Information Administration, 2010). Energy consumption is loosely correlated with gross national product and climate. USA consumes 25% of the world's energy with a share of global GDP at 24% and a share of the world population at 5%. The most significant growth of energy consumption is currently taking place in China, where it holds about 20% of the world population around 1.3 billion people in 2008, and its GDP contributes 7% of the world's total and has been growing at 5.5% per annum over the last 30 years since China's economic reform in 1978. Similarly, as shown in Figure 4, the primary energy consumption in China also grows at an annual speed of 5.5% from 1978 to 2008 (National Bureau of Statistics, 2009).

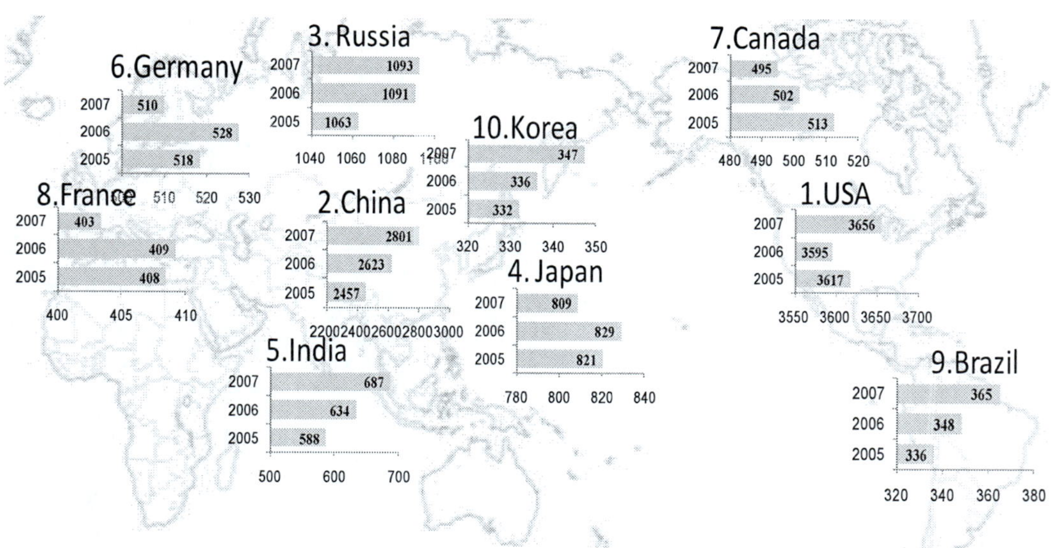

Unit: million TCE

note: no. before country name denotes the sequence in energy consumption

Figure 3. Top 10 largest energy consumption countries in the world.

Secondly, to grasp the status of energy consumption of each industry especially the transportation sector, Figure 5 describes the change of sectoral primary energy consumption from 1995 to 2007. It is found that energy use from transport sector increased from 59 MTCE in 1995 to 206 MTCE in 2007. Although the volume of energy consumed by transport ranks the third, just behind industry and household sectors in China, the growth of energy use in transport sector is the fastest, with an average rate of 11% per annum during 1995-2007.

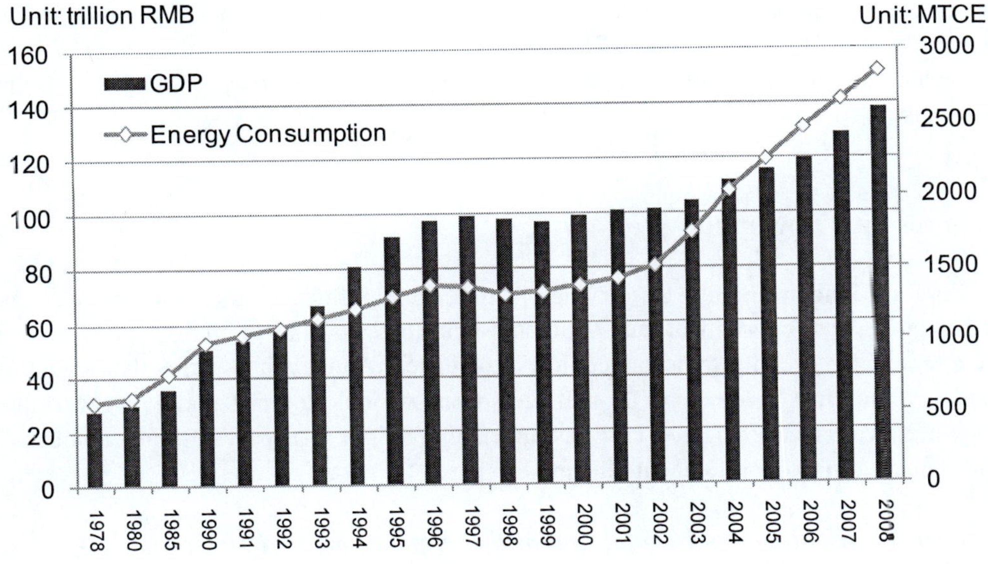

Figure 4. GDP and primary energy consumption in China (1979-2008).

Note: GDP is in 2000 price.

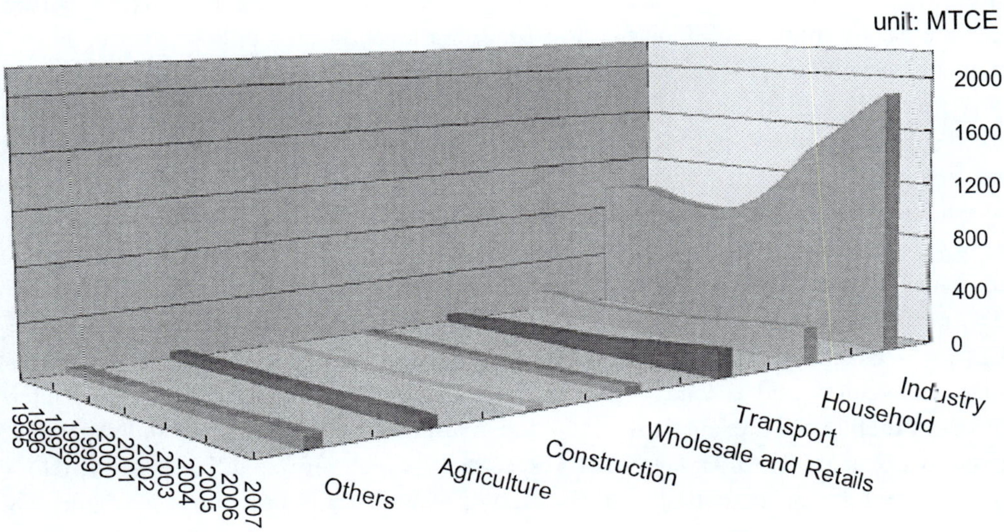

Figure 5. Energy consumption by sector (1995-2007).

2. Air Pollutant Emissions fron Transport Sector and Evaluation Methodology

In this study, transport sector is composed of two general types. One is inter-city passenger and freight transport system, which consists of railway, road, waterway and airway modes. Another is urban transport, which includes private car, bus, trolley, taxi and rail modes with a further division by fuels. Ferry is dropped out due to data limitation. The inter-modal system enables transport through different modes, used in sequence and/or as alternatives. The problems are concerned in this session with 1) the characteristics of inter-city and urban transport volume and modal share; 2) the external impacts of transport development on non-renewable energy use, and methods for evaluation the transport oriented air pollutant emissions.

2.1 Inter-city Transport

For the understanding of driving factors that influence transport related energy consumption and consequent air pollutant emissions so that preferred policies toward environmentally friendly transport could be identified. Annual time series data of inter-city passenger and freight transport is collected from various sources mainly concerning the following two aspects: a) inter-city passenger and freight transport's turnover volume and modal share; b) factors influencing the modal share.

2.1.1 Turnover volume and modal split of passenger and freight transport

Data about turnover volume and modal split of passenger and freight transport is taken from such as "China Statistical Yearbook" (NBS, 2000-2009), "Comprehensive Statistical Data and Materials on 50 Years of New China: 1949-1998" (NBS, 1999). It is found that the total turnover volume of both passenger and freight transport has been increasing significantly during the past 30 years. As described in Figure 6, since 1978, the passenger mobility has been increasing from 174 billion p-km in 1978 to a level of 2320 billion p-km in 2008, with an annual growth rate around 9.0%. Freight turnover volume also grew at an annual speed of 8.5% from 940 billion ton-km in 1978 reaching 9955 billion ton-km in 2007. However, the modal split differs between passenger and freight transport. More specifically, except for road and airway, the shares of railway and waterway passenger transport kept decreasing during the whole period. Waterway made the least contribution to the total mobility growth, while the biggest proportion changed in turn between railway and road. Before the 1990s, although the contribution of road mode to total passenger turnover volume increased from 30% to 47%, railway transport still shared the biggest percentage of total mobility over 50%. While after the 1990s, with the large amount of investment on road and the popularization of motorized vehicle, the fast increase in road transport played a dominant role in inter-city passenger mobility. For example, in 2004, the investment on road, railway and waterway was about 337, 65 and 39 billion yuan RMB respectively, which reflects different priority for each mode's development and partially explains road's 54% share of total turnover volume. On the other hand, the inter-city freight transport relies heavily on water, rail and road modes whereas the share of airway transport has remained negligible in its comparison. Before the mid-1980s, railway is the dominant way for transporting freight with its share above 45%. After that waterway took over and became the most important freight mode. The modal share growth rates have varied, with road services growing at a much faster rate than waterways'.

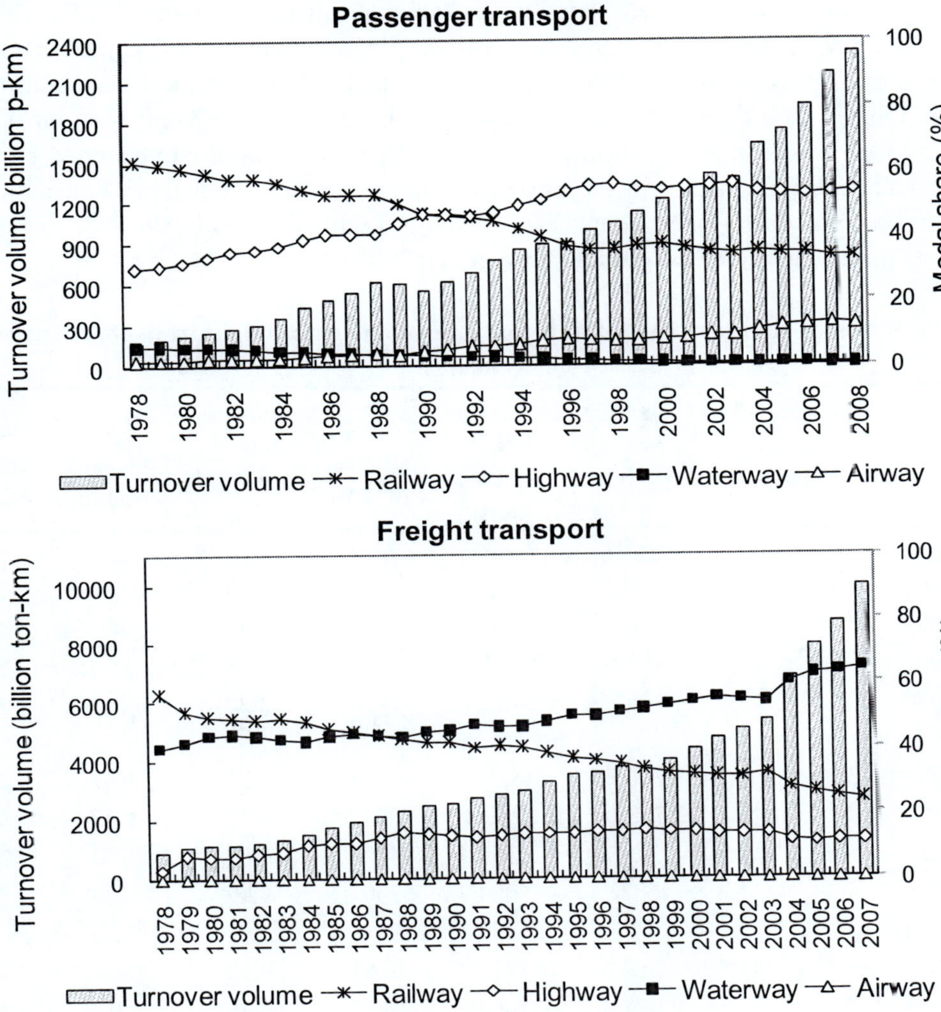

Figure 6. Turnover volume and modal split of passenger and freight transport.

2.1.2 Factors influencing modal share

The general goals concerned in this study are to switch the transport structure toward a more environmentally friendly one so that fuel consumption and air pollutant emissions could be reduced. For realizing the goals, affecting factors are identified in terms of specific parameters such as passenger capacity or freight volume, traffic network extension rate by mode and fuel tax rate, etc. Through the different planning preference and investment for each mode's network extension, government can induce the development of high capacity and low environmental load modes such as railway and waterway. Furthermore, as an important instrument for restraining the growth of fuel demand and associated air pollution, fuel tax has been adopted in many countries and proved effective (Sterner, 2007). The price of fossil fuel is the summation of industrial production cost and taxation. The latter is expected to reduce the fossil fuel depletion and encourage the usage of clean energy. In this session, transport modal share is assumed to be determined by the cumulated passenger capacity, traffic network length and fuel cost per transport unit, etc.

Annual data about these major factors affecting modal share from 1978 to 2008 is taken from "Yearbook of China Transportation & Communication" (Yearbook House of China Transportation & Communication, 1986-2005), "Price Yearbook of China" (Editorial Department of Price Yearbook of China, 1997-2005) and "China Statistical Yearbook" (NBS, 2009). Table 1 shows the data in some selected years. Although different types of fuel are used by each transport mode, in this analysis, fuel consumption is all converted based on thermal amount to the same type of gasoline due to the limitation in fuel price data. The fuel conversion parameter is taken from "China's Energy Yearbook" (NBS, 2005). In Table 1, the gasoline price figures are changed into the constant 2000 price.

Table 1. Modal characteristics of inter-city passenger and freight transport in selected years

	Modal share (%)		Transport capacity		Traffic network length		Fuel intensity *l*/(ton-km)	Gasoline price (yuan/*l*)
	passenger	freight	passenger (million pop)	freight (million tons)	length (10^3 km)	annual growth rate (%)		
Railway								
1980	60.6	49.6	922.0	1112.8	53.3	1.43	0.0066	2.1
1990	46.4	41.5	957.1	1506.8	57.8	0.27	0.0055	1.5
2000	37.0	31.7	1050.7	1785.8	68.7	1.93	0.0035	3.7
2008	33.5	23.9	1461.9	3142.4	79.7	2.20	0.0030	2.6
Road								
1980	32.0	6.6	2300.0	3820.5	883.3	3.21	0.13	2.1
1990	46.6	13.1	7600.0	7240.4	1028.3	4.24	0.12	1.5
2000	54.3	14.0	13290.0	10388.1	1402.7	5.67	0.11	3.7
2008	53.8	11.4	26821.1	16394.3	3730.8	4.09	0.11	2.6
Waterway								
1980	5.7	43.8	264.4	383.8	108.5	0.18	0.011	2.1
1990	2.9	45.3	272.3	706.9	109.2	0.46	0.009	1.5
2000	0.8	54.2	193.9	994.4	119.3	0.14	0.008	3.7
2008	0.3	64.6	203.3	2223.0	122.8	-0.59	0.007	2.6
Airway								
1980	1.7	0.01	3.4	0.1	114.1	10.4	0.62	2.1
1990	4.1	0.03	16.6	0.4	340.4	11.4	0.58	1.5
2000	7.9	0.11	67.2	2.0	994.5	-0.4	0.55	3.7
2008	12.4	0.12	192.5	4.0	1341.7	3.6	0.43	2.6

In order to find out the contribution of each factor to both passenger and freight transport modal share in the past decades and provide parameters for future projection, the following regression model is adopted for analysis.

$$\text{MS}_{i,t} = C_i + a_{1,i}\text{CAP}_{i,t} + a_{2,i}\text{NET}_{i,t} + a_{3,i}\text{FC}_{i,t} + a_{4,i}T \quad (1)$$

where, subscript i and t denotes transport mode and year; MS is modal share in percentage; CAP is cumulated passenger capacity or freight volume, which is assumed to have a positive effect on modal share, the more passengers or freight are carried the larger modal share will be; NET is traffic network length. Due to the expansion of accessibility and attractiveness for vehicles, network extension is regarded as one of the most important factors encouraging the mode development. FC is fuel cost per transport unit, calculated by multiplying fuel price with fuel consumption per transport unit. Through which, policy measures by imposing taxes can be evaluated. The more fuel cost per transport unit is the less attractiveness of mode will be; A time trend variable, T, is introduced to reflect an upward or downward trend in modal share change that cannot be explained by the other explanatory variables, which takes value 1 through 31 from 1978 to 2008; C is a constant. To allow for excessive multicollinearity among the independent variables, stepwise regression estimation is used although it is recognized that this has very serious technical pitfalls but is easier to interpret than say factor analysis.

Table 2 shows the results. The adjusted R^2 in each case is close to 1.0, which indicates the linear combination of explanatory variables is sufficient and reliable for modal share estimation.

Table 2. Determinants of modal share in inter-city passenger and freight transport

	Modal share (%)		Transport capacity		Traffic network length		Fuel intensity	Gasoline price
	passenger	freight	passenger (million pop)	freight (million tons)	length (10^3 km)	annual growth rate (%)	l/(ton-km)	(yuan/l)
Railway								
1980	60.6	49.6	922.0	1112.8	53.3	1.43	0.0066	2.1
1990	46.4	41.5	957.1	1506.8	57.8	0.27	0.0055	1.5
2000	37.0	31.7	1050.7	1785.8	68.7	1.93	0.0035	3.7
2008	33.5	23.9	1461.9	3142.4	79.7	2.20	0.0030	2.6
Road								
1980	32.0	6.6	2300.0	3820.5	883.3	3.21	0.13	2.1
1990	46.6	13.1	7600.0	7240.4	1028.3	4.24	0.12	1.5
2000	54.3	14.0	13290.0	10388.1	1402.7	5.67	0.11	3.7
2008	53.8	11.4	26821.1	16394.3	3730.8	4.09	0.11	2.6
Waterway								
1980	5.7	43.8	264.4	383.8	108.5	0.18	0.011	2.1
1990	2.9	45.3	272.3	706.9	109.2	0.46	0.009	1.5
2000	0.8	54.2	193.9	994.4	119.3	0.14	0.008	3.7
2008	0.3	64.6	203.3	2223.0	122.8	-0.59	0.007	2.6
Airway								
1980	1.7	0.01	3.4	0.1	114.1	10.4	0.62	2.1
1990	4.1	0.03	16.6	0.4	340.4	11.4	0.58	1.5
2000	7.9	0.11	67.2	2.0	994.5	-0.4	0.55	3.7
2008	12.4	0.12	192.5	4.0	1341.7	3.6	0.43	2.6

2.2 Urban Transport

2.2.1 Private car stock and public transport

As another major component of transport sector, urban transport plays an important role in stimulating economic development, improving quality of life of citizens, and inducing urban landscape formation, etc. However, in most of the developing countries, due to the insufficient supply of public transit system and road infrastructure when comparing with mobility demand, a series of problems such as booming of personal vehicles, traffic congestion, long travel time and air pollution, etc.

As illustrated in Figure 7, the urban transport system in China is generally composed of private car and public transport mode, which can be further divided into bus, trolley, taxi and rail. Ferry is dropped out due to data limitation.

As pointed out in many literatures, motor vehicle stock is strongly related to economic level (International Energy Agency, 2001; World Business Council for Sustainable Development, 2004; Das and Parkh, 2004). Thus when estimating the private car stock and associated air pollutant emissions, this study defines a country-region-province hierarchical structure for China, noting that the whole country is composed of regions and region consists of provinces. Following the regional clusters proposed by the Development Research Center of State Council of China (2005), we divide China's 31 provinces into 8 regions according to the economic homogeneity (Figure 8).

All the socioeconomic data is collected at the provincial level with urban and rural division. In detail, annual data of GDP, per capita income, population, inter-province and intra-province migration etc. are taken from "China Statistical Yearbook" (National Bureau of Statistics, 1996-2006) and "Tabulation on the Population Census of the People's Republic of China" (National Bureau of Statistics, 1993, 2002). By choosing area, population and GDP as indices, defining each one as the proportion of regional value to China's total, Table 3 illustrates the general characteristics of regional socio-economy. It is remarkable that the regional development is significantly unbalanced. The eastern and coastal regions (NE, NC, EC and SC) cover only 20% of whole country's territory. However, they accommodate 43% of total population and create 63% of GDP in China. On the contrary, the western and inland regions (SW and NW) share over 50% of area while only 24% of population concentrates there and 14% of GDP are created.

Table 3. China's regional characteristics (unit: %)

	NE	NC	EC	SC	MYR	MCR	SW	NW	China
Area	9.3	4.5	2.5	4.0	20.6	8.5	16.4	34.2	100
Population	8.5	14.4	10.7	9.5	14.9	18.5	19.1	4.4	100
GDP	9.8	18.5	19.9	14.3	10.5	13.2	10.9	2.9	100

Note: the shares of population and GDP are the average value during 1995-2005.

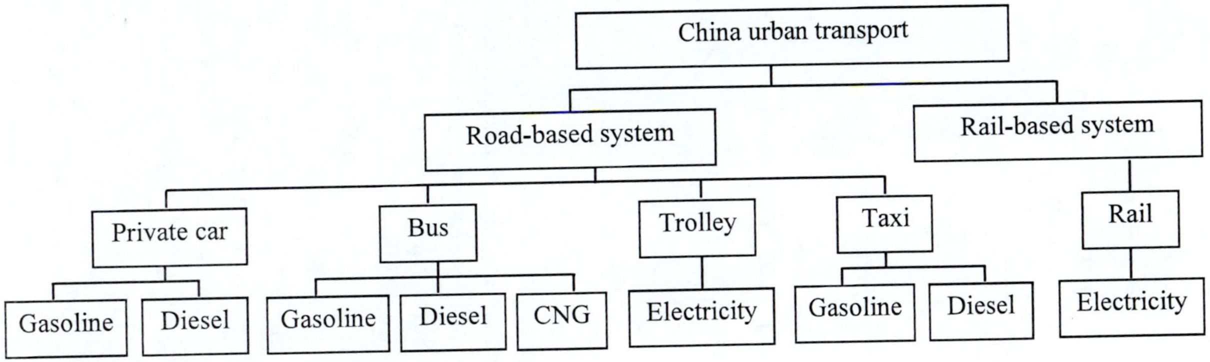

Figure 7. Hierarchy of China urban passenger transport system.

NE: Liaoning, Jilin, Heilongjiang

NC: Beijing, Tianjin, Hebei, Shandong

EC: Shanghai, Jiangsu, Zhejiang

SC: Fujian, Guangdong, Hainan

MYR: Shanxi, Inner Mongolia, Henan, Shaanxi

MCR: Anhui, Jiangxi, Hubei, Hunan

SW: Guangxi, Chongqing, Sichuan, Guizhou, Yunnan

NW: Tibet, Gansu, Qinghai, Ningxia, Xinjiang

Figure 8. Eight regions in China.

Data on private car such as private car stock per hundred households, budget share of private car in annual per capita expenditure and car price index etc. are compiled from "China Urban Life and Price Yearbook" (Department of Urban Society and Economic Statistics, 1996-2006) and "China Yearbook of Rural Household Survey" (Department of Rural Society and Economic Statistics, 1996-2006). Figure 9 shows the private car stock and per capita income in urban and rural areas of China's 8 regions in 2000. Overall the private car stock in urban and coastal regions is higher than that in rural and inland areas, with the level far below the world average. For example, in the wealthiest region (south coast urban area, SC-u), the private car stock per 100 people was 0.35. While in the least developed region (northwest rural area, NW-r), it was only 0.04. By comparison, there are over 70 private cars per 100 people in US, 40 in Japan and 35-50 in Europe.

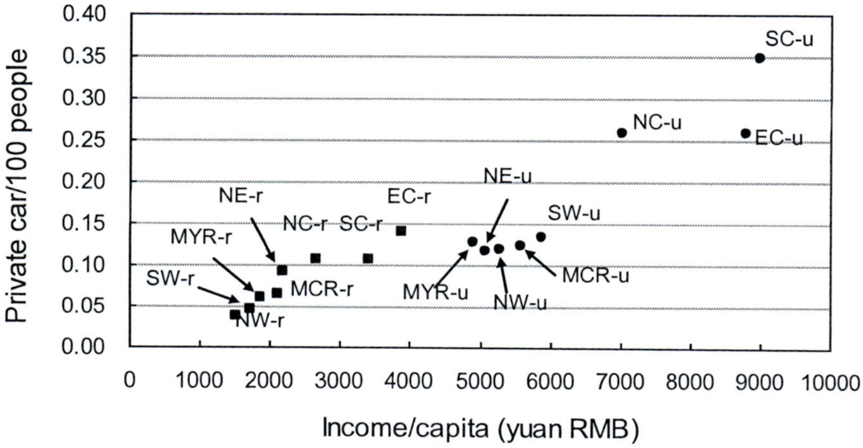

Figure 9. Private car stock and per capita income in 8 regions of China in 2000.

Note: "u" and "r" refers to urban and rural area respectively.

Data on urban public transport is collected from 31 provinces from 2000 to 2006 including vehicle number, passengers transported by rail, annual distance traveled by vehicles (DTV), average trip length traveled by passenger (DTP) and fuel intensity etc. are mainly taken from "China Statistical Yearbook" (National Bureau of Statistics, 2001-2007), "China Urban Construction Statistical Yearbook" (Department of Finance, Ministry of Construction, PRC, 2004, 2007), published reports and papers. Table 4 shows the characteristics of four urban public transport modes with further division by fuel type. Overall during the period of 2000-2006, the growth rate of vehicular number by each mode have varied, with bus increasing at a fastest rate of 5.7% per annum, taxi growing slowly at a rate of 1.9% per annum, while trolley decreasing at 5% per annum. The passenger amount transported by rail increased fast in such a short period at an annual rate of 25% touching 2332 million in 2006, which indicates a rapid development of rail mode in urban transport.

Table 4. Characteristics of road and rail based modes

	Vehicle No. ('000)		Proportion (%)		DTV ('000 km)		FI	
	2000	2006	2000	2006	2000	2006	2000	2006
Bus	222	310						
Gasoline			7	13	52	50	0.44	0.42
Diesel			83	75	52	50	0.32	0.31
CNG			10	12	52	50	0.38	0.38
Trolley	3.4	2.5						
Electricity			-	-	34	32	1.4	1.4
Taxi	826	929						
Gasoline			99.9	97.5	26	24	0.16	0.15
Diesel			0.1	2.5	26	24	0.16	0.15
	Passenger transported (million pop)		DTP (km/passenger)		FI	
	2000	2006	2000	2006	2000	2006
Rail	607	2332	9	9	0.05	0.05

Note: The unit of FI is (L or m^3)/km for bus and taxi, kWh/km for trolley, and kWh/p-km for rail.

2.2.2 Energy consumption and emission factor of urban transport

Firstly, in order to estimate the energy consumption and related air pollutant emissions from private car, two more factors besides private car stock are needed—annual distance traveled by private car (DTC) and fuel intensity (FI).

Since there is no published statistical data on DTC in China, we follow the scenario analysis results conducted by Wang et al. (2007), and assume that the more private car stock is in a region the less intensively each car is used. Noting that DTC in those developed countries with high private car stock ranges mainly from 10000 km to 18000 km (International Road Federation, 2005), we define the DTC in China will converge to 14000 km per annum and propose a logistic function to measure DTC change with the increase of private car stock (Fig 10).

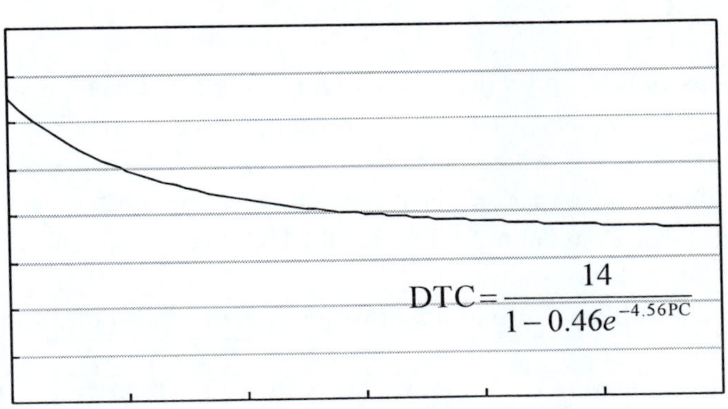

Figure 10. Relationship between DTC and per capita private car stock.

Fuel intensity is usually determined by many local factors such as fuel quality, engine type, driver's habit and road condition etc. However, due to the data limitation we assume private car in each region has the same FI. In recent years, China government has been implementing intensive transport policies with general targets aimed at encouraging the use of light vehicles and improving fuel economy. Notably, the Standardization Administration of the People's Republic of China (2004) issued the "Maximum Limits of Fuel consumption for Passenger Cars" in 2004, stressing the realization of target that the fuel consumption per 100 km of car and light duty vehicles could decrease 15% in 2006-2010 than the current level. In this context and taking the existing literature as a base, we project the FI and proportion of gasoline and diesel cars (Table 5).

Table 5. Fuel intensity and proportion of private car classified by fuel type

	2000	2005	2010	2020	2030
Proportion (%)					
Gasoline car	99.91	97.48	95.02	92.49	90.01
Diesel car	0.09	2.52	4.98	7.51	9.99
Fuel intensity (*l*/100 km)					
Gasoline car	16.2	15.56	14.94	14.35	13.78
Diesel car	16.2	15.56	14.94	14.35	13.78

In this study, the methodology provided by the Intergovernmental Panel on Climate Change (1997) is adopted for the assessment of air pollutant emissions from private car use mainly including two aspects. One is emissions of CO_2 and CH_4, two major greenhouse gases of global impact. The other is emissions of CO, Non-Methane Volatile Organic Compounds (NMVOC), NO_x and SO_2 mainly of local effect.

Eq. (2) is used for estimating vehicular emissions in province *i*.

$$\begin{aligned} \mathrm{EC}_{m,i,t} &= \sum_n \mathrm{PC}_{n,i,t} \cdot \mathrm{POP}_{n,i,t} \cdot S_{m,t} \cdot \mathrm{DTC}_{n,i,t} \cdot \mathrm{FI}_{m,t} \cdot a_m \\ \mathrm{EM}_{q,i,t} &= \sum_m \mathrm{EC}_{m,i,t} \cdot b_{q,m} \end{aligned} \tag{2}$$

where, subscripts *m* and *q* denote fuel type and air pollutants respectively; EC is energy consumption; EM is vehicular emissions; *S* is proportion of private car using different fuel; *a* is heat conversion factor, which takes value as 3.2×10^7 J/*l* for gasoline and 3.6×10^7 J/*l* for diesel; *b* is emission factor per unit heat generation in road transport sector, which is complied from the revised IPCC guidelines for national GHG inventories. The detailed values are listed in Table 6.

Table 6. Emission factor for each air pollutant in kg/TJ

	CO_2	CH_4	CO	NMVOC	NO_x	SO_2
Gasoline car	68607	20	8000	1500	600	4.65E-05
Diesel car	73326	5	1000	200	800	1.41E-04

Secondly, the following equation is proposed to estimate the energy consumption (EC) of both road- and rail-based urban public transport systems.

$$\begin{aligned} &\text{Road - based mode:} \quad EC_{m,t} = \sum_j VN_{m,j,t} \cdot DTV_{m,j,t} \cdot FI_{m,j,t} \cdot a_j \\ &\text{Rail - based mode:} \quad EC_{rail,t} = PN_{rail,t} \cdot DTP_{rail,t} \cdot FI_{rail,t} \cdot a_j \end{aligned} \tag{3}$$

where subscript m, j represent road-based mode and fuel type respectively. VN is vehicle number. DTV is the annual distance traveled by vehicle. PN is passenger transported by rail. DTP is average trip length traveled by passenger. FI is the fuel intensity, which is defined as fuel/km for road-based mode, and fuel/passenger-km for rail mode. a is heat conversion factor, which takes value as 3.9×10^7 J/m^3 for CNG, and 3.6×10^6 J/kWh for electricity.

3. Strategies and Policies for Energy Conservation and Emission Control

Saving energy and mitigating air pollution are the major tasks and challenges for transport sector since a long time ago in China. Table 7 lists out the relevant strategies, policies, regulations and criterions for the mitigation of energy consumption and emissions since the 1980s. Generally they could be divided into three levels: national law, national plan and policy, transport sectoral policy and criterion. In detail, there are two major laws issued from the 1990s, in which energy conservation in transportation sector is especially emphasized as an independent chapter in Law on Energy Conservation. In the level of national plans and policies, we find that more and more intensive transport policies have been implemented in China with the general targets aiming at encouraging the development of light vehicles and public transport, and improving the fuel economy. Moreover, the Eleventh Five-Year Plan for China's transport sector goes further and stresses that GHG emission should be controlled. As far as infrastructure network is concerned, the overall plan sets forth the priorities as 1) to accelerate the development of railway transportation; 2) to improve road network; and 3) to further develop waterway transport. In 2006-2010, totally 0.092 million km of railway, 2.3 million km of highway and 0.13 million km of inland waterway system will be constructed. Moreover, because transport management in China is organized separately according to different transport modes, each one works according to its own policies, operational strategies, and allocates projects and builds its own network of relationships across other authorized departments. For instance, road and water transport are administered under the Ministry of Communications; railways fall under the authority of the Ministry of Railways; civil aviation is administered under the Civil Aviation Administration. Each transport administration department issues plans, rules and criterions such as standards for the scrapping of motor vehicles issued by State Planning Commission, State Economic & Trade Commission in 1997, and policy on energy conservation technologies in railway issued by Ministry of Railways in 1999, etc.

As another important economic policy instrument, fuel tax is recognized as an important macroeconomic measure controlling transportation development, which has been approved efficient in many countries. In China, although it has not yet been implemented so far, central government has already enacted fuel tax legislation in 1999. And the possible tax rate is supposed to range from 30% to 50% at the beginning of policy implementation (Zhang, 2006). Therefore, when assessing the effect of transport policies on air pollutant emissions mitigation, fuel tax must be taken into account as an important policy option. Generally, the transport system in China lacks comprehensiveness, while an integrated transport management administrative system is still inchoate.

Table 7. Transport policies concerning energy conservation and emission reduction

Level	Issue year	Issuer	Law / Policy / Plan / Regulation
National Law	1998	Order of the Chairman of the People's Republic of China	Law on Energy Conservation
	2005	The National People's Congress of the People's Republic of China	Law of the People's Republic of China on Renewable Energy
National plan and policy	2004	National Development and Reform Commission	Mid- and Long-term Specific Plan on Energy Conservation
	2005	General Office of the State Council of PRC	Opinion of Encouraging the Development of Energy Saving and Environmental Protection Type Low-emission Vehicle
	2006	National Development and Reform Commission	National Eleventh Five-Year Plan
	2006	National Development and Reform Commission	Technology and Policy Outline of Energy Conservation
	2006	State Council of PRC	Decision of the State Council on Reinforcecement of Energy Conservation Works
	2007	State Council of PRC	Synthetic Plan on energy conservation and emission reduction
Transport sectoral policy	1986	Ministry of Railways	Provisional Detailed Rules on the Management of Energy Conservation in Railways
	1992	Ministry of Transportation	Rules for the Publication of Energy Conservation Products of Automobiles and Vessels
	1997	State Planning Commission, State Economic & Trade Commission, etc.	Standards for the scrapping of motor vehicles
	1999	Ministry of Railways	Policy on Energy Conservation Technologies in Railway
	2000	General Administration of Quality and Technology supervision	Limits for Automobile Emission of Pollutants and Their Testing Method
	2000	Ministry of Transportation	Detailed Rules on the Implementation of Energy Conservation Law in the Transportation Industries
	2002	Ministry of Railways	Criterions on Energy Conservation in Railway Engineering Design

4. FUTURE PROJECTION OF AIR POLLUTANT EMISSIONS AND POLICY CONTROL UP TO 2030

In this session, we firstly propose a SD model for the projection of energy consumption and reduction potential of air pollutant emissions and CO_2 emissions of global impact from inter-city passenger and freight transport. Secondly, we adopt a linear expenditure model and regression analysis for projection of urban private car stock and public transport growth and related air pollutant emissions.

4.1 A System Dynamics Model of Air Pollutant Emissions from Inter-city Transport

As an important tool supporting policy experiments, system dynamics (SD) methodology can not only arrange and describe the complicated connections among each element at different levels, but also deal with dynamic process with feedback in complex systems. A number of studies have applied this approach related to environment, such as analysis of GHG reduction and global warming (Vrat et al, 1993), sustainability evaluation (Chen et al. 2006) and environmental planning and management (Guneralp and Barlas, 2003).

Figure 11 shows the relationships among the items of goal, policy and resulting effects in the model. Dynamics of the model are determined by the feedback loops in the causal loop diagram. Each arrow of the causal loop diagram indicates the influence of one element on the other. The influence is considered positive (+) if an increase in one element causes an increase in another, or negative (-) in the opposite case. Broken line indicates an incentive relationship. The base year of projection is 2000, for which the later years' data can be used for model validation. While 2030 is set as the target year by taking into account the reliability of projection results.

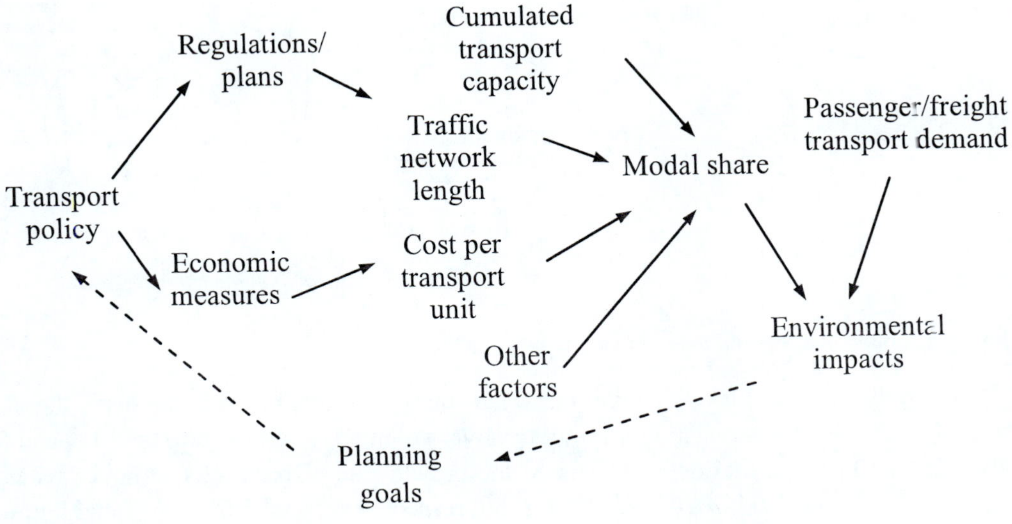

Figure 11. Causal loop diagram of the system dynamics model for inter-city transport.

In calculation, a flow diagram is used to show the detailed physical and information flow in the system dynamics model (Figure 12). The general goals concerned in this study are to reduce the fuel consumption and to mitigate air pollution. For realizing the goals, policies are identified in terms of specific parameters such as traffic network extension rate by mode and fuel tax rate, etc. Through the different preference and investment for each mode's network extension, government can change the transport structure toward a more environmentally friendly one. Furthermore, as an important instrument for restraining the growth of fuel demand and associated CO_2 emissions, fuel tax has been adopted in many countries and proved effective (Sterner, 2007). The price of fossil fuel is the summation of industrial cost and taxation. The latter is expected to reduce the fossil fuel depletion and encourage the usage of clean energy. In the system dynamics model, transport modal share is assumed to be determined by the cumulated passenger capacity or freight volume, traffic network length and fuel cost per transport unit, etc. As for the projection of passenger and freight transport demand, usually it can be estimated by two methods. One is to extrapolate the historical trend by regression analysis, where the passenger turnover volume (p-km) or freight volume (ton-km) is regarded as explained variable, and population, per capita GDP, vehicle number, etc. are adopted as explaining variables (Haldenbilen, 2006). The other way is to estimate the transport demand according to the national development strategy and plan (Piattelli et al., 2002). In this study, the latter method is used by setting the growth rate of total passenger turnover volume according to the National Eleventh Five-Year Plan.

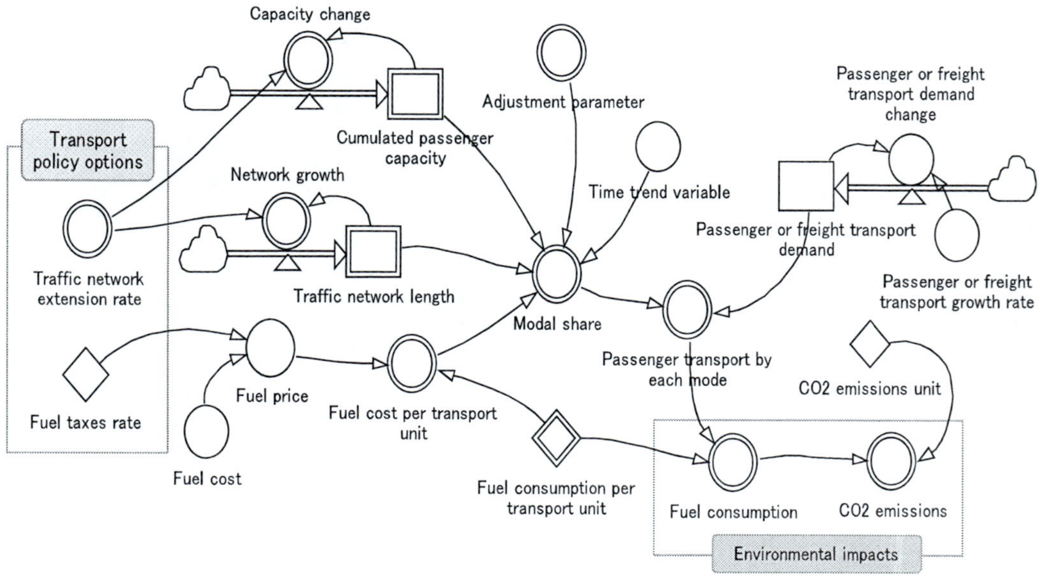

Figure 12. Flow diagram of the system dynamics model.

According to the Eleventh Five-Year Plan, during the period of 2006-2010, the annual growth rate of total passenger and freight turnover volume are supposed to be 9.1% and 5.6% (Wu, 2006). In the scenarios, we assumed the growth rate of passenger turnover volume at 10% and 11% in 2011-2020 and 2021-2030 respectively, and 5.6% for freight turnover volume during 2011-2030. Based on the analysis of past transport policies and existing studies, three scenarios are established (Figure 13). Generally, the business and usual (BAU) scenario extrapolates historical trends and assumes the traffic network will increase at the average rate for 2000 to 2005. The middle control scenario is based mainly on the Eleventh Five-Year Plan embracing policies emphasizing energy conservation and air pollution mitigation by accelerating railway and waterway construction and introducing fuel tax rate at 45%. In the high control scenario, more intensive policies are implemented concerning the environment, notably that the government will invest heavily in infrastructure construction giving more priority to railway and waterway while slowing the pace of roads and airway expansion Furthermore, a fuel tax rate to 50% is assumed.

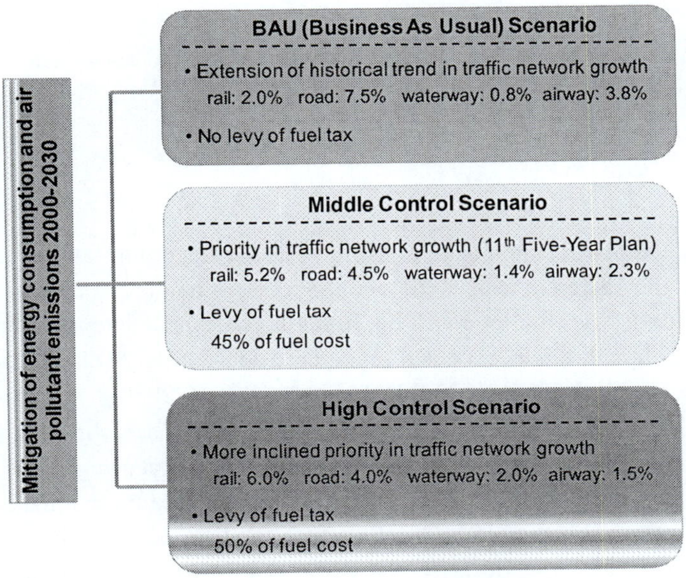

Figure 13. Air pollution mitigation scenarios for inter-city transport.

4.2 Model of Urban Transport Development and Emissions

In Chinese statistics, private car represents the licensed four wheeler passenger vehicle with capacity of 2-8 persons. It only serves for personal use and can be further classified into gasoline and diesel cars. Instead of extrapolating the historical trend to the future, in this study we treat private car as a consumable good and adopt a linear expenditure system, notably the Stone-Geary model for projection. The base year is 2000, with 2005 used for validation, and 2030 is set for the target year. As described in Eq. (4), consumer aims at getting a maximum utility from consuming goods with the restriction of limited income.

$$\begin{aligned} Max: \quad & U_{n,i,t} = \alpha_{n,i,t} \ln \mathrm{NPC}_{n,i,t} + (1-\alpha_{n,i,t}) \ln(G_{n,i,t} - \beta) \\ s.t. \quad & Y_{n,i,t} = p_{\mathrm{NPC}n,i,t} \cdot \mathrm{NPC}_{n,i,t} + p_{Gn,i,t} \cdot G_{n,i,t} \end{aligned} \qquad (4)$$

where subscript n refers to either urban or rural areas; i and t denote the province and year; U is consumer utility; NPC is newly purchased private cars per capita; G is a generic good representing all other consumer goods; α is the budget share of private cars in annual per capita expenditure; β is the subsistence level of the good. As suggested by Meyer et al. (2007), the definition of β is same as that of poverty, which is set as 50% of one unit generic good for empirical analysis. Y is per capita disposable income; p_{NPC} and p_G are the prices of a private car and of the generic good respectively.

Using Lagrangean and first order conditions to solve Eq. (4), we obtain

$$\mathrm{NPC}_{n,i,t} = \frac{\alpha_{n,i,t}}{2-\alpha_{n,i,t}} \cdot \frac{Y_{n,i,t}}{p_{\mathrm{NPC}n,i,t}} \tag{5}$$

The private car stock per capita is a function of NPC and cars owned the previous year after subtraction of scrapped cars.

$$\mathrm{PC}_{n,i,t} = \mathrm{NPC}_{n,i,t} + \mathrm{PC}_{n,i,t-1} \cdot (1-\sigma_{n,i,t-1}) \tag{6}$$

where PC is the private car stock per capita; σ is the scrapping rate for private cars, which may vary with the different stages of economic development between regions: the less developed a region is, the smaller σ will be. In 2002, the annual scrapping rate for private cars was only 4.6%, although this is expected to increase substantially (Zhang and Zong, 2008). In the projection, we set it at 5% in 2000-2015 and 6% in 2016-2030.

By inputting the PC, α and Y data for the urban and rural areas in each province for 1999 and 2000, we obtain the price of a private car in 2000 as a starting value and then use the average growth rate in car prices during 1995-2000 to estimate future change up to 2030. The share of private cars in household budgets is quite small in 2000, with a national average of 1.5% in urban and 0.8% in rural areas. However, with increased income levels, the budget share is expected to increase. In this analysis, we set this annual growth in the budget share at 8% between 2000 and 2030 in both urban and rural areas.

For the projection of the private car stock, the input of two driving socio-economic factors is required: per capita income and population. We follow the works of International Institute for Applied Systems Analysis (2003a,b) for modeling. Since they observed the historical patterns of demographic process in China and extrapolated into the future by assuming continuing trends without consideration of migration under specific economic development, we use their results at provincial level as a base, and adopt a gravity model-inspired approach to model marginal changes in population distribution due to the relative income changes between provinces, and between urban and rural areas. Furthermore, this study highlights the modeling of regional economic inequality through a economic convergence estimation. Here, the issue whether or not economic convergence will take place across China's provinces has attracted wide attentions, and been examined by a number of research (Yao and Zhang, 2001; Brun et al., 2002; Sakamoto and Islam, 2008). As discussed by Han et al. (2007), the total GDP inequality in China does not exhibit a significant trend of divergence during the period of 1978-2004. More importantly, in order to achieve sustainable development, China government has been making great efforts to reduce the disparities between regions and urban-rural areas. The implementations of "Develop-the-West Strategy" and "Eleventh Five-Year Plan" in recent years are well known actions. The promotion of harmonious regional development has been emphasized as one of the targets in future development. Therefore, when estimating the future economic development, regional economic convergence must be taken into account. For the detailed evaluation methods of per capita income and population, please refer to the work of Han and Hayashi (2008).

Regarding the estimation of public transport development and their air emissions, as pointed out in many literatures, motor vehicle number is strongly correlated with economic level. Thus we use an elasticity-based method to project the public use vehicle (VN) as shown in Eq. (7).

$$\mathrm{VN}_{i,m,t} = a_{i,m} + b_{i,m}\mathrm{GDP}_{i,t} \qquad (7)$$

where, subscript i, m, and t denote province, urban public transport mode and year. We assume the vehicles for public use such as trolley, taxi and bus are strongly related to GDP change (in billion yuan RMB).

The estimation methods of air pollutant emissions from urban private car and public transport are same with Eq. (2) and Eq. (3).

4.3 Model Results

Firstly, Figure 14 demonstrates the scenarios in terms of energy consumption and CO_2 emissions in 2000-2030. Generally, energy consumption and CO_2 emissions in the inter-city passenger and freight transport will keep increasing up to 2030 even if some proper policies are implemented. However, the effects of suggested policies on energy conservation and CO_2 mitigation are noticeable. Specifically, under the BAU scenario, historical patterns are assumed to continue in the future and there is no specific policies stress CO_2 mitigation. Energy consumption in 2030 is supposed to reach 62500 Peta Joules, which is more than 13 times the amount in 2000. CO_2 emissions also record the same sharp increase and amount to 4289 million tons in 2030. In middle and high control scenarios, emphases are laid on the adjustment of transport structure and introduction of fuel tax. With more and more inclined priorities for the development of railway and waterway and higher fuel tax rate, in 2030, energy demand under the middle and high control scenarios is projected to be 38600 and 33500 Peta Joules respectively. Similarly, the corresponding CO_2 emissions is also expected to be reduced by 38% and 46% of that in BAU scenario.

In this study, policy parameters include traffic network extension rate (railway, highway and waterway) and fuel tax rate. However, some parameters may have greater effect on CO_2 mitigation, while others may be less effective. Thus, we conduct sensitivity analysis of each policy parameter by Eq. (8).

$$S = \frac{\Delta \mathrm{EM}_t}{\mathrm{EM}_t} \Big/ \frac{\Delta X_t}{X_t} \qquad (8)$$

where, S is the sensitivity of a specific parameter in year t; EM is CO_2 emissions; X is policy parameter influencing CO_2 emissions; ΔEM and ΔX are the increments or decrements of CO_2 emissions (EM) and parameter (X) respectively.

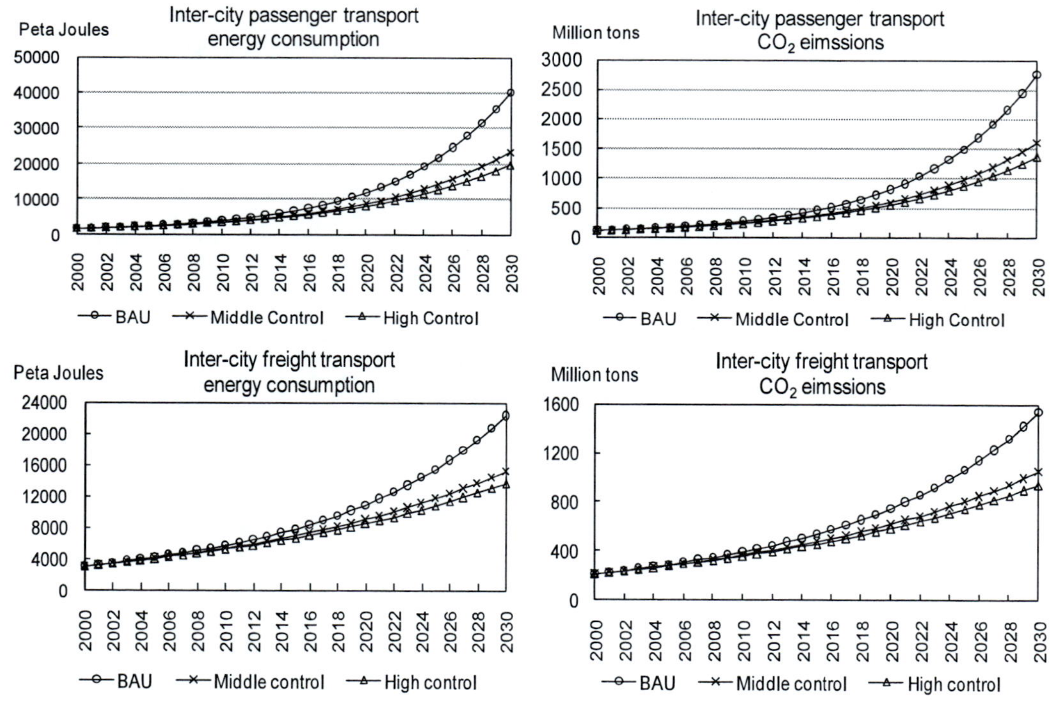

Figure 14. Energy consumption and CO_2 emissions in inter-city transport 2000-2030.

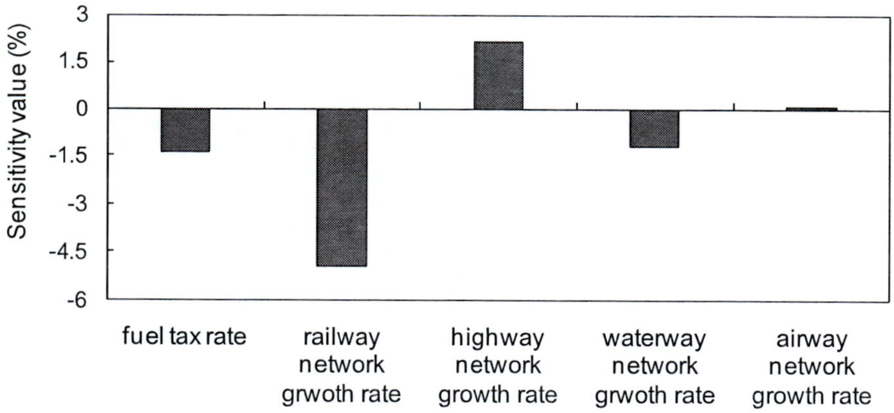

Figure 15. Sensitivity analysis of policy parameters.

In calculation, we assume each parameter will increase by 10% every five years during the period of 2001-2030. By using Eq. (8), four sensitivity values for each parameter will be obtained. Then, we use their average value to represent the general sensitivity of the parameter to CO_2 emissions. Results are shown in Figure 15, where except the road and airway network's growth rate is sensitive to the increase of CO_2 emissions, while the other three parameters are sensitive to the reduction of CO_2 emissions. Moreover, when considering the absolute value of sensitivity, growth rate of railway network, growth rate of road network, fuel tax rate and growth rate of waterway network are the most sensitive three parameters in order. Their values are 5.0%, 2.1%, 1.4% and 1.2% respectively. The growth rate of airway network is least sensitive to CO_2 emissions in the system dynamics model, with its sensitivity only valued by 0.1%.

Secondly, Figure 16 illustrates the projected trends of air pollutant emissions from urban transport till 2030 by mode-wise. Generally, energy consumption in China urban transport sector will keep increasing. During the period 2000-2030, the total volume of CO_2, CH_4, CO, NMVOC, NO_x and SO_2 emissions will increase sharply with annual average growth rates of 16.5%, 15.9%, 15.8%, 15.8%, 16.6% and 17.6% respectively. By the end of 2030, the volumes of these emissions will be 631.4, 0.2, 79.2, 15.5, 8.5 and 0.8 million tons respectively. The massive growth of the urban private car stock will be the major reason for this, for which urban private car contributes to 80-90% of total air pollution from urban transport. Further, due to the imbalance in growth between urban and rural areas, the annual increase in vehicle emissions will also be much larger in the urban areas. In 2000, cities accounted for around 54% of all the emissions, but by the end of 2030 this proportion will rise to 83%.

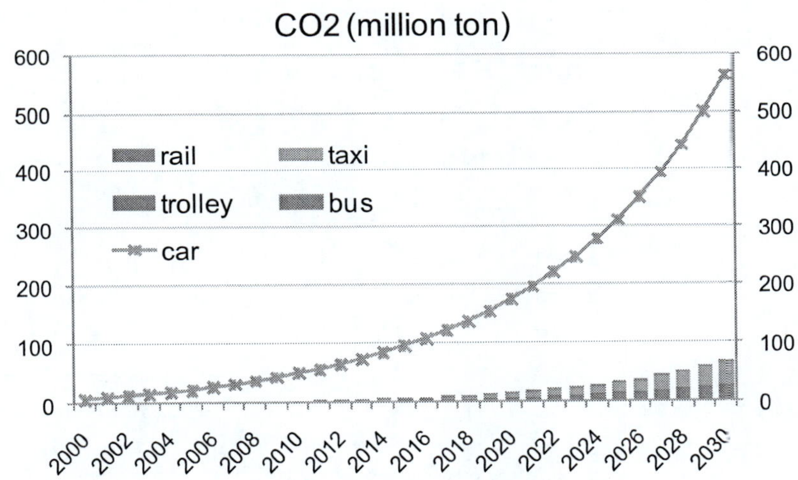

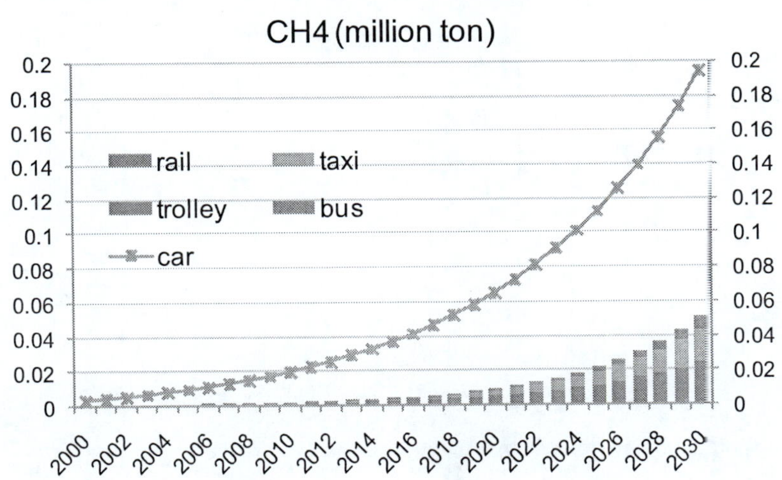

Figure 16 (Continued).

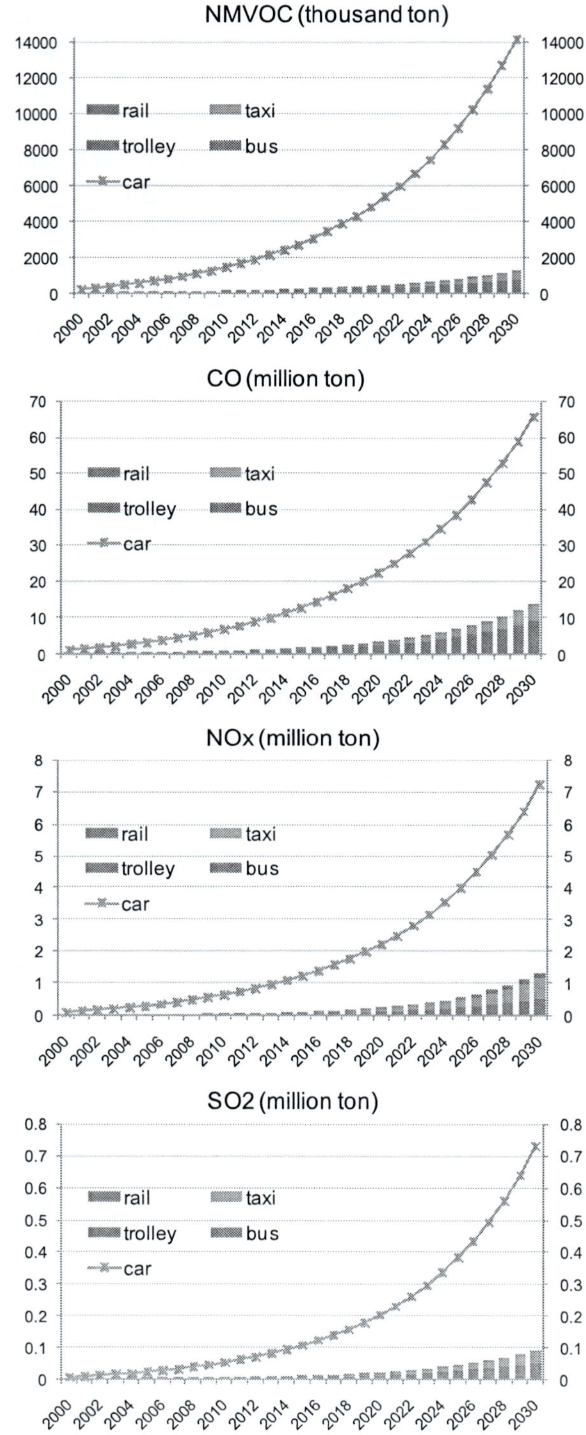

Figure 16. Air pollutant emissions from urban transport.

CONCLUSION

In the context of global warming, studies on how to mitigate the non-renewable energy use and air pollutions in transport sector have been paid wide attentions especially in the recent years and near future. For developing countries, it is increasingly important to investigate appropriate strategies and policies to slow down the increase in local air pollutant and green house gas emissions. This paper chooses the inter-city and urban transport in China as a case, and applies a system dynamics model for policy assessment and CO_2 mitigation scenario analysis for inter-city passenger and freight transport, and an expenditure model of urban private car and public transport. In contrast with previous works, we have highlighted the modeling of the macroscopic driving forces behind transport development, especially against the background of the current phase of rapid socio-economic transition. Factors usually ignored in the existing literature, such as the economic imparities among regions, population migration, policy influences and their interactions, were investigated to allow a appropriate projections of transport demand and corresponding emissions of local and global effects. The results indicate that the total transport amount, but also the volume of related pollutant emissions will shoot up to considerably higher levels in the near future if recent behavioral trends and the present technical aspects of private car use persist. Despite the introduction of stricter controls on private car purchase and pollutant emissions, China will come under much greater pressure to cut back on emissions.

However, a huge mitigation potential exists in the current inter-city transport in China. In 2030, a figure for CO_2 emissions as low as 2300 million tons could be achieved if policies in the high control scenario are well implemented. Comparing to the case without any specific policies stressing CO_2 mitigation, the reduction of CO_2 emissions will range from 38% to 46% under those scenarios with policy controls. Among all the policy options examined in this paper, sensitivity analysis suggests that accelerating the development of railway network is the most effective option to reduce CO_2 emissions the in China's inter-city passenger transport. Slowing down road network extension and levying fuel taxes, and accelerating waterway network extension are also significant and useful policies for air pollutant mitigation.

REFERENCES

Brun, J.F., Combes, J.L., Renard, M.F. (2002). Are there spillover effects between coastal and non coastal regions in China. *China Economic Review*, 13, 161-169.

Cai, H., Xie, S. (2007). Estimation of vehicular emission inventories in China from 1980 to 2005. *Atmospheric Environment*, 41, 8963-8979.

Chen, H.B., X. Wang, and B.H. Mao. (2006). Sustainability Assessment of Urban Transport Policy-A Case Study of Beijing, *China. Journal of Transportation Systems Engineering and Information Technology*, 6 (3), 100-112.

Das, A., Parikh, J. (2004). *Transport scenarios in two metropolitan cities in India: Delhi and Mumbai.* Energy Conversion and Management, 45, 2603-2625.

Department of Rural Society and Economic Statistics. (1996-2006). *China Yearbook of Rural Household Survey*. China statistics press, Beijing (in Chinese).

Department of Urban Society and Economic Statistics. (1996-2006). *China Urban Life and Price Yearbook.* China statistics press, Beijing (in Chinese).

Department of Finance, Ministry of Construction, PRC. (2004, 2007*). China Urban Construction Statistical Yearbook. China Architecture & Building Press*, Beijing, (in Chinese).

Development Research Center of State Council of China. (2005). Strategies and policies for regional harmonious development. Available on *http://www.drc.gov.cn/view.asp?doc_ID=030955* (in Chinese).

Editorial Department of Price Yearbook of China. (1997-2005). *Price Yearbook of China*,. Beijing Huazheng Press, Beijing (in Chinese).

Energy Information Administration. (2010). *International Energy Statistics*. Available on *http://tonto.eia.doe.gov/cfapps/ipdbproject/IEDIndex3.cfm?tid=44&pid=44&aid=2*

Guneralp, B., Barlas, Y. (2003). Dynamic modeling of a shallow fresh water lake for ecological and economic sustainability. *Ecological Modeling*, 167, 115-138.

Haldenbilen, S. (2006). Fuel price determination in transportation sector using predicted energy and transport demand. *Energy Policy*, 34, 3078-3086.

Han, J., Zhou, X., Imura, H. (2007). A disparity analysis of regional GDP and CO2 emissions in China based on Theil and shift-share method. *Environmental Science*, 20(6), 449-460.

Han, J., and Hayashi, Y. (2008). Assessment of private car stock and its environmental impacts in China from 2000 to 2020. *Transportation Research Part D: Transport and Environment*, 13(7), 471-478 (Elsevier, SCI journal impact factor: 1.319).

He, K. (2005). Oil consumption and CO2 emissions in China's road transport: current status, future trends, and policy implications. *Energy Policy*, 33, 1499-507.

Huang, C., C.H. Chen, B.Y. Wang, Y. Dai, J. Zhao, and H.K. Wang. (2005). Urban Travel Modal Split and Its Impact on Energy and Environment. *Journal of Highway and Transportation Research and Development,* 22 (11), 163-166 (in Chinese with English abstract).

International Energy Agency. (2001*). Rapid motorization in the largest countries in Asia: implication for oil, carbon dioxide and transportation.* Available on *http://www.iea.org/textbase/papers/2001/rapmot.pdf.*

International Institute for Applied Systems Analysis. (2003a). *Scenario analysis on urbanization and rural- urban migration in China.* Available on *http://www.iiasa.ac.at/Publications/Documents/ IR-03-036.pdf.*

International Institute for Applied Systems Analysis. (2003b*). Regional population projections for China.* Available on *http://www.iiasa.ac.at/Admin/ PUB/Documents/IR-03-042.pdf.*

International Panel on Climate Change. (2007). IPCC Fourth Assessment Report, Working Group III Report "*Mitigation of Climate Change*". Available on *http://www.ipcc.ch/.*

International Road Federation. (2005). *World road statistics 2005*. Geneva.

Kuhns, H.D., Mazzoleni, C., Moosmuller, H., Nikolicb, D., Keislara, R.E., Barbera, P. W., Li, Z., Etyemezian, V., Watsona, J.G. (2004). Remote sensing of PM, NO, CO and HC emission factors for on-road gasoline and diesel engine vehicles in Las Vegas, NV. *Science of the Total environment*, 322, 123-137.

Meyer, I., Leimbach, M., Jaeger, C.C. (2007). International passenger transport and climate change: a sector analysis in car demand and associated CO2 emissions from 2000 to 2050. *Energy Policy*, 35, 6332–6345.

National Bureau of Statistics. (2000-2009). *China Statistical Yearbook*. China Statistics Press, Beijing.

National Bureau of Statistics. (1999). *Comprehensive Statistical Data and Materials on 50 Years of New China: 1949-1998.* China Statistics Press, Beijing. (in Chinese).

National Bureau of Statistic. (2005). *China's Energy Yearbook*. China Statistics Press, Beijing (in Chinese).

Piattelli, M.L., Cuneo, M.A., Bianchi, N.P., Soncin, G. (2002). The control of goods transportation growth by modal share re-planning: the role of a carbon tax. *System Dynamics Review,* 18, 47-69.

Sakamoto, H., Islam, N. (2008). Convergence across Chinese provinces: an analysis using Markov transition matrix. *China Economic Review*, 19, 66-79.

Standardization Administration of the People's Republic of China. (2004). *Maximum limits of fuel consumption for passenger cars*. Standards Press of China (in Chinese).

Sterner, T. (2007). Fuel taxes: An important instrument for climate policy. *Energy Policy*, 35, 3194-3202.

Vrat, P., Gupta, Y.K., Gupta, A. (1993). *A system dynamic study of global warming*. Proceedings of the xviith national system conference. Kanpur.

Wang, C., Cai, W., Lu, X., Chen, J. (2007). CO2 mitigation scenarios in China's road transport sector. *Energy Conversion and Management*, 48, 2110-2118.

World Business Council for Sustainable Development. (2004). *Mobility 2030: meeting the challenges to sustainability.* The Sustainable Mobility Project, Switzerland.

World Conference on Transport Research Society and Institute for Transport Policy Studies. (2004). *Urban transport and the environment: an international perspective*. Amsterdam, Elsevier.

Wu, X. (2006). Discussions on Some Issues of the Eleventh Five-Year Plan for Transport of China. *Journal of Transportation System Engineering and Information Technology,* 6, 1-5.

Yao, S.J., Zhang, Z.Y. (2001). On regional inequality and diverging clubs: a case study of contemporary China. *Journal of Comparative Economics*, 29, 466-484.

Yearbook House of China Transportation & Communication. (1986-2005). *Yearbook of China Transportation & Communication*. China Agriculture Press, Beijing (in Chinese).

Zhang, G.L., Zong G. (2008). Using scrapping rate to predict the demands of the car industry. *Auto Industry Research,* 6, 10-11 (in Chinese).

Zhang, T.J. (2006). Research on quantitative analysis: case study of change from fuel fare to fuel tax. *Marketing Research*, 12, 27-30.

In: Pollution in China
Editor: Michael I. Chang
ISBN: 978-1-61122-022-3
©2011 Nova Science Publishers, Inc.

Chapter 5

INDOOR AIR POLLUTANTS IN CHINA: LEVELS, SOURCES, AND RISKS OF VOCS AND PAHS

***Mitsuhiro Kojima*[1], *Lizhong Zhu*[2], *and Takeshi Ohura*[3*]**

[1]Graduate School of Nutritional and Environmental Sciences,
University of Shizuoka, 52-1 Yada, Shizuoka 422-8526, Japan
[2]Department of Environmental Science, Zhejiang University,
Hangzhou, Zhejiang 310028, China
[3]Faculty of Agriculture, Meijo University,
1-501 Shiogamaguchi, Nagoya 468-8502, Japan

ABSTRACT

In order to investigate indoor air quality for China, much of surveys for various air pollutants have been conducted. This chapter is focused on the pollution of aromatic volatile organic compounds (VOCs) and carcinogenic polycyclic aromatic hydrocarbons (PAHs) in indoor air in Hangzhou in China. The surveys were conducted in indoor microenvironments (living room, bedroom, and kitchen) and outdoors, which were compared to the corresponding data obtained in Shizuoka, one of urban cities in Japan. Comparing the contributions and relationships among those pollutants, the significant differences of related emission sources were clarified between both countries; throughout the samplings, the indoor and outdoor concentrations of many of the targeted VOCs (benzene, toluene, ethylbenzene, xylenes, and trimethylbenzenes) in China were significantly higher than those in Japan. The indoor concentrations of VOCs in Japan were somewhat consistent with those outdoors, whereas those in China tended to be higher than those outdoors. The concentrations of indoor PAHs in China and Japan also showed the similar trends of the case of VOCs; the level in China was extremely higher (>10-times) than that in Japan. Finally, the lifetime cancer risks estimated from unit risks and geometric mean indoor concentrations of carcinogenic VOCs were 24×10^{-5} in China and 2.6×10^{-5} in Japan. For PAHs, Toxicity potencies of PAHs in residential air of China were much higher than that in Japan. These estimations indicate that the exposure risks for air pollutants in China could be problem of deep concern.

* Corresponding author: Takeshi Ohura, E-mail: ohura@meijo-u.ac.jp.

1. INTRODUCTION

As a result of the rapid economic growth for the two decades since the initiation of economic reforms in 1978, China has been experiencing a rapid urbanization created by the history's largest flow of rural-urban migration in the world (Zhang and Song, 2003). Since 1990, decoration and refurbishment have become popular in urban civil buildings in China. In an airtight indoor environment designed for energy-saving purpose, harmful compounds emitted from decorating and refurbishing-materials tend to accumulate due to low air exchange rate (Wang et al., 2004). Indeed, air pollution levels have increased the past decade due to the increases in motor vehicles, urban construction, and industrial combustion (Chan and Yao, 2008).

Volatile organic compounds (VOCs) are dominant contributor in the air pollutants, which can induce a range of chronic adverse health effects such as irritation to the eyes, mucous membranes, skin, and respiratory tract, and effects on the nervous system, cancer, and liver and kidney toxicity (Delfino et al., 2003; Jones, 1999; Nielsen et al., 2007). Aromatic hydrocarbons, such as benzene, toluene, ethylbenzene, and the isomeric xylenes (BTEX), are abundant in the indoor environment. In particular, benzene and toluene could be pollutants that are considerably investigated the ambient behaviors and exposure risks because of its high toxicity and/or high contents in the air (Boström et al., 1994; Low et al. 1988; Hein et al., 2008; Holian, 1996; Kume et al., 2008; Ohura et al., 2006). The primary emitters of BTEX to air include industrial sources, mobile sources, various consumer and household products, and building and construction materials (Elbir et al., 2007; Guo et al., 2004a,b; Kim et al., 2002; Liu et al., 2008; Loh et al., 2006; Ohura et al., 2006). For example, mobile sources are largely responsible for the presence of benzene, toluene, ethylbenzene xylenes, and 1,3-butadiene in the air (Elbir et al., 2007; Guo et al., 2004a,b; Liu et al., 2008). For indoor sources, insulation, smoking, solvents and coating medium furniture, and room decoration were also the potent emission sources (Polzin et al., 2007; Finlayson-Pitts and Pitts, Jr. 2000; Na et al., 2004).

Polycyclic aromatic hydrocarbons (PAHs) associated with particles are carcinogenic and mutagenic compounds produced by incomplete combustion of organic substances (Maliszewska-Kordybach, 1999; Mastral and Callen, 2000). Anthropogenic sources of PAHs include the combustion of materials for energy supply such as coal, oil, gas, and wood (Maliszewska-Kordybach, 1999; Mastral and Callen, 2000). In particular, plants such as coke and aluminum production, residential heating, and power generation as stationary and mobile sources comprise the majority of human-produced PAH sources (Mastral and Callen, 2000). Natural sources of PAHs include volcanic activity and forest fire (Maliszewska-Kordybach, 1999). Indoor emission sources of PAHs include smoking, cooking, heating, and furniture (Liu et al., 2001; Ohura et al., 2004).

Humans spend most of their lives indoors. Therefore, any evaluation of the human health risks caused by exposure to air pollution requires a detailed understanding of pollutant levels indoors. In addition, this understanding must include an estimation of the contribution of outdoor pollutants to indoor pollution. In the present study, we examine the distribution and concentration of particle matter and associated PAHs in indoor and outdoor air in two industrial cities, Fuji and Shimizu, in Japan. On the basis of these data, we characterized the seasonal (summer vs. winter) behavior and relationships of the pollutants. We also present an environmental health risk assessment for indoor carcinogenic PAH concentrations.

2. Material and Methods

2.1. Sampling Sites

Air sampling was conducted in summer and winter in Hangzhou, China and Shizuoka, Japan. Sampling in Hangzhou was conducted in August 2006 for summer and January 2007 for winter and sampling in Shizuoka was conducted from August to September 2006 for summer and from January to March 2007 for winter. Hangzhou (latitude 30°16'N, longitude 120°12'E; Figure 1) in China has a temperate climate (average temperature of 16.2°C), 6,600,000 inhabitants, and is a large commercial city. Shizuoka (latitude 35°09'N, longitude 138°42'E; Figure 1) in Japan has a similar climate (average temperature of 16.3°C), about 700,000 inhabitants, and is typical of a commercial city in Japan.

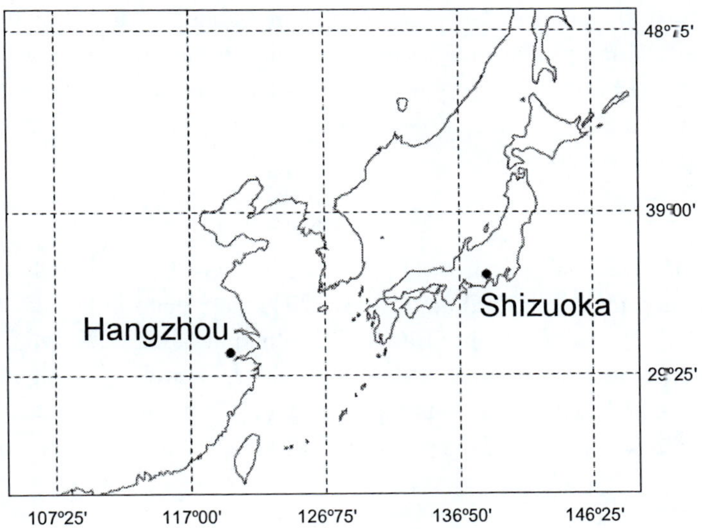

Figure 1. Locations of surveyed cities.

In China, 14 houses were sampled in summer and winter, with seven houses being sampled both in summer and winter. All houses surveyed is present in urban center of each city. Information on house characteristics was obtained from a self-administered questionnaire of the inhabitants. In Japan, 30 and 27 houses were sampled in summer and winter respectively. Among them, 27 houses were sampled both in summer and winter. Characteristic differences in house structures were observed between China and Japan; half of the houses in Japan were constructed with wood (53%) and the remainder was constructed with steel (47%), whereas most of the houses in China were built using concrete. In addition, the inner walls of most houses in Japan were covered by wallpaper, whereas most in China were painted. Two people (7%) and one person (4%) in Japan and five (36%) and four people (29%) in China smoked during the monitoring period rein summer and winter respectively.

Although smoking is well known to be significant indoor source of VOCs, our data relating to smoking were very limited, and thus we abandoned further analysis of this source.

2.2. Sampling Methods

VOCs

Air samples were collected in the living room, kitchen, a bedroom and outside for each house in both countries. The outdoor sampler was placed away from exhaust ducts and heat sources of the house and protected from rain and direct solar irradiation. In addition, most of the outdoor samplers were placed in backyards of the houses to prevent aspiration and adhesion of dust due to human activity. Field blanks of each of the four samples were deployed on each sampling day.

Ambient VOCs were collected by passive samplers packed with activated charcoal (Sibata Chemicals, Co. Ltd, Tokyo, Japan) for 24 h in Japan and by active samplers (Tenax-TA, SUPELCO, USA) to a personal pump (DDY-1.5, Xingyu, China) in China at a flow rate of 0.5 L/min for 30 min. One of the reasons that different sampling methods were used for Japan and China is to avoid the saturation of compounds on the samplers because VOC concentrations in China are expected to be much higher than in Japan. Also, accuracies of those methods selected in each country have been already confirmed (Ohura et al., 2006; Li et al., 2009). After sampling was complete, the samplers were stored at –45°C in sealed aluminum bags until extraction.

PAHs

Gaseous PAHs were adsorbed using XAD-2 (Supelco, USA), which was cleaned with dichloromethane and methanol until no peak of PAHs was found in HPLC. Particulate PAHs were collected with 25 mm glass fiber filters (GF, Whatman, England), which were preheated at 400°C for 6 h to remove organic compounds before sampling. Both XAD-2 tubes and filters were connected with a mini-pump (Xingyu, China for Chinese samples; Sibata Scientific Technology, Japan for Japanese samples) for air sampling. The sampling flow rate was 1.0 L/min.

2.3. Extraction and Analysis

VOCs

The adsorbent (activated charcoal) from the passive samplers was transferred to a glass test tube containing 2.00 mL of distilled carbon disulfide, containing benzene-*d*6 and toluene-d_8 as internal standards. The tube was shaken mechanically (115 times/min) at room temperature for 10 min, and this was followed by centrifugation for 10 min, 3000 rpm. The supernatant (1.00 mL) was then transferred into a vial for gas chromatography/mass spectrometry (GC/MS). VOC samples were analyzed using an HP 6890 gas chromatograph with a 5972A mass selective detector (GC-MSD, Hewlett Packard, Palo Alto, CA, USA). Other GC/MS conditions and quantification methods for VOCs have been described elsewhere (Ohura et al., 2006). On the other hand, analysis of samples collected by the active sampler in China was conducted using a thermal desorber (Fuli 9700, China) interfaced with a gas chromatograph equipped with a flame ionization detector (Fuli 9790, China). Chromatographic separation was achieved using a polyethylene glycol capillary column (20 m, 0.25 mm). The sample tube was loaded in the desorber, which was heated to 250°C for 10 min. Nitrogen gas was then passed through the tube and samples were quickly injected into the chromatographic column. The temperatures of the injector and detector were equally set as 260°C. The oven temperature of the gas chromatograph was initially held at 60°C for 5 min, then raised to 150°C at a rate of 5°C/min and then held for another 2 min. The 9 compounds (benzene, toluene, ethylbenzene, p-xylene, m-xylene, o-xylene, 1,3,5-trimethylbenzene, 1,2,4-trimethylbenzene, and 1,2,3-trimethylbenzene) were targeted in this study, which were of the highest purity available (Wako Pure Chemicals, Osaka Japan or IERM, China). The detail information related calibration curves were described in elsewhere (Amagai et al., 2002; Li et al., 2009).

PAHs

The sampled XAD-2 were poured into a 25 mL glass stoppered tube containing 20 mL mixture of dichloromethane and acetonitrile (3/2 = v/v). The glass fiber filters were cut into pieces, placed in a 25 mL glass tube containing 10 mL dichloromethane. The samples were sonicated for 30 min while the water in the ultrasonic bath was replaced frequently in order to prevent overheating. Subsequently, 10 mL extracts of XAD-2 and 5 mL extracts of glass fiber filters were transferred into new glass stoppered tubes. Extracts with 30 μL dimethyl sulfoxide (DMSO) were evaporated under a gentle flow of nitrogen gas at room temperature and then 970 μL acetonitrile were added.

The samples from Hangzhou sites were analyzed for the following 16 PAHs: naphthalene (Nap), acenaphthylene (Acy), acenaphthene (Ace), fluorene (Flu), phenanthrene (Phe), anthracene (Ant), fluoranthene (Fluor), pyrene (Py), benzo[*a*]anthracene (BaA), chrysene (Chry), benzo[*b*]fluoranthrene (BbF), benzo[k]fluoranthrene (BkF), benzo[a]pyrene (BaP), dibenzo[*a,h*]anthracene (DBahA), benzo[*ghi*]perylene (BghiP), indeno[1,2,3-cd]pyrene (IP). The air samples from Shizuoka sites were analyzed for the following 8 PAHs: BaA, Chry, BbF, BkF, BaP, DBahA, BghiP, IP. PAHs sampled in residential air of Hangzhou were determined by HPLC (Agilent, USA) containing a Lichrospher PAH column (250 × 4.6 mm, Agilent, USA), a fluorescence detector and an ultraviolet detector (determine ACY only). PAHs sampled in residential air of Shizuoka were determined by HPLC (Hitachi, Japan) containing a Wakosil-II PAH column (250 × 4.6 mm, Wako Chemicals, Japan) and a fluorescence detector (Hitachi, Japan). Thus, only the 8 PAHs (BaA, Chry, BbF, BkF, BaP, DBahA, BghiP, IP) simultaneously determined in both countries were compared to discuss the difference of residential air PAH pollutions in the two countries.

3. Results and Discussion

3.1. VOC Concentrations

Throughout sampling in China and Japan, all VOCs targeted were detected above the detection limits at each monitoring site. The highest geometric mean concentration (mean) of total VOCs indoors in summer in Japan was observed in kitchens, followed by living rooms and bedrooms, but the order in winter was living rooms, kitchens, bedrooms (Table 1). In China, on the other hand, the order for indoors was bedrooms, living rooms, kitchens for summer, and living rooms, bedrooms, kitchens for winter. Among indoor microenvironments, the cleaner sites were bedrooms in Japan but kitchens in China. Among the target VOC concentrations in summer in Japan, toluene was the most abundant at all surveyed sites, followed by ethylbenzene, m-xylene, and 1,2,4-trimethylbenzene (1,2,4-TMB) in all indoor microenvironments. However, the order in winter in Japan was different from that in summer, with the concentrations of ethylbenzene tending to be lower than those of m-xylene and 1,2,4-TMB. In China, toluene was most abundant in indoor microenvironments as for Japan, whereas the concentrations of benzene were higher than ethylbenzene and m-xylene concentrations.

Of the outdoor VOC concentrations, the total concentrations in Japan were somewhat consistent with or lower than those indoors. On the other hand, the total concentrations in China were a few times lower than those indoors. The total VOC concentrations at each monitoring site in China were several times higher than the corresponding site concentrations in Japan; 3.4 (outdoors) to 9.4 (bedrooms) times higher in summer, and 2.0 (kitchens and outdoors) to 2.7 (bedrooms) times higher in winter. This suggests the presence of significantly stronger indoor and/or outdoor emission sources of VOCs in China.

Table 1. Total VOC concentrations in indoor microenvironment and outdoor in China and Japan

	China			Japan			
Summer	mean	min	max	mean	min	max	China/Japan
living room	84.32	29.97	792.10	12.14	2.21	55.15	6.9
bedroom	86.14	28.40	427.92	9.17	1.66	18.03	9.4
kitchen	62.92	25.78	383.56	12.35	1.93	56.75	5.1
outdoor	35.00	6.88	302.72	10.43	1.13	22.39	3.4
Winter	mean	min	max	mean	min	max	
living room	59.83	17.31	164.91	24.52	8.12	78.93	2.4
bedroom	54.13	14.82	209.99	19.92	5.66	48.88	2.7
kitchen	46.75	14.13	138.84	23.77	5.12	85.54	2.0
outdoor	27.61	13.02	57.91	13.60	1.16	43.63	2.0

Concerning the effects of VOC concentrations on smoking activity, benzene and toluene concentrations in living room in China were compared. The comparison showed that these compounds in houses with smokers were approximately 1.6 times higher than in houses without smokers, although there were no statistically significant differences between them.

3.2. PAH Concentrations

Table 2. Indoor PAH concentrations in summer and winter (ng/m^3)

Summer

	China			Japan		
PAHs	Mean	Max	Min	Mean	Max	Min
BaA	35	2.60E+02	4.4	0.22	1.9	0.011
CHRY	7.3	35	2.2	0.7	5	0.032
BbF	1.4	8.3	0.17	1.2	5.1	0.031
BkF	0.69	3.7	0.072	0.47	3	0.01
BaP	0.61	2.1	0.11	0.65	5	0.017
DBahA	0.33	2.1	nd	0.13	0.74	nd
BghiP	1.2	3.6	0.15	0.94	5.5	0.026
IP	0.65	2.1	nd	0.9	5.4	0.018
∑PAHs	47	320	7.1	5.2	32	0.15

Winter

	China			Japan		
PAHs	Mean	Max	Min	Mean	Max	Min
BaA	12	40	4.2	0.38	1.9	0.016
CHRY	6.1	13	1.7	0.86	5.3	0.029
BbF	5.6	14	0.57	1.4	8.9	0.055
BkF	2.4	59	0.32	0.44	2.5	0.019
BaP	4.4	12	0.52	0.85	4.7	0.039
DBahA	0.69	3.2	0.18	0.14	0.7	nd
BghiP	4.3	9.4	0.58	1.1	5.8	0.053
IP	4.9	13	nd	0.96	5.1	0.039
∑PAHs	40	160	8.1	6.1	35	0.26

PAH concentrations in indoor air of two countries are presented in Table 2. Here, the indoor data in China and Japan showed the results of representative indoor, living room (Ohura et al., 2005). In summer, total concentrations of 8 PAHs (∑PAHs) ranged from 7.1 to 320 ng/m^3 and 0.15 to 32 ng/m^3, with an average of 47 ng/m^3 and 5.2 ng/m^3 in residential air of China and Japan, respectively. In winter, the corresponding concentrations varied from 8.1 to 160 ng/m^3 and 0.26 to 35 ng/m^3, with an average of 40 ng/m^3 and 6.1 ng/m^3 in the two countries, respectively. The results indicated that ∑PAHs in China were significantly higher than that in Japan. Among the 8 PAHs, concentrations of 4-ring PAHs (BaA and Chry) were much higher in residential air of China, whereas concentrations of 5 to 6-ring PAHs were similar in residential air of two countries. BaA was observed to be the most abundant in residential air of China, which could be emitted from cooking practice (Zhu and Wang, 2003). The concentrations of BaA accounted for 62.0% to 80.0% and 24.4% to 52.0% of ∑PAHs in summer and winter, respectively. In residential air of Shizuoka, BbF was the most abundant, accounting for 16.2% to 20.5% and 21.3% to 25.6% of ∑PAHs in summer and winter, respectively. Most of high molecular weight (MW) PAHs are mutagenic and carcinogenic to humans (Caricchia et al., 1999; Maliszewska-Kordybach, 1999). BaP concentrations were observed ranging from 0.11 to 12 ng/m^3 and 0.017 to 5.0 ng/m^3 in residential air of Hangzhou and Shizuoka, respectively, which indicated that the concentrations were much higher than those in Houston (0.0027-1.1 ng/m^3) and Elizabeth (0.0055-0.23 ng/m^3) (Naumova et al., 2002).

The average summer/winter ratios of PAH concentrations were calculated to assess the seasonal variation of individual PAHs (Fig. 2). Concentrations of 4-ring PAHs were higher in summer than those in winter in China, with corresponding summer/winter ratios higher than 1.0. However, the opposite case was found in Japan. In addition, the ratios for 5 to 6-ring PAHs were both lower than 1.0, especially in China. This phenomenon was reported to relate to photodegradation to related compounds in summer, and reductions in combustion temperature.

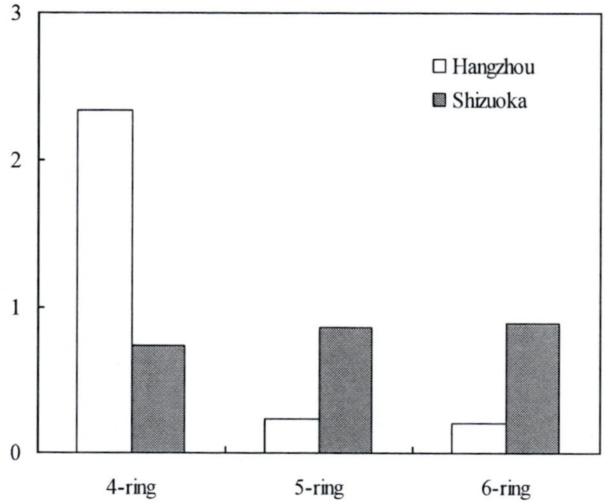

Figure 2. Average summer/winter Ratios of PAH concentrations.

3.3. Profiles

By evaluating correlations of air pollutants among the sampling sites, it is possible to determine their behaviors and predict their emission sources. First, we investigated the relationships between indoor and outdoor concentrations of VOCs (benzene and toluene) in each country (Table 3). For each VOC, significant correlations in Japan were observed between concentrations in living rooms and outdoors (Table 3). However, there were no significant correlations between the concentrations for indoor and outdoor. Comparing the benzene concentrations at each monitoring site in Japan, the outdoor concentrations were slightly higher than concentrations in living rooms (see Table 1). Among indoor rooms, the living room concentrations were consistent with or slightly higher than concentrations in other indoor rooms. That is, indoor (living room and kitchen) benzene concentrations could be strongly affected by outdoor pollution, and benzene could easily spread from room to room. Such indoor concentrations in China were frequently higher than outdoor concentrations (Table 1). Consequently, in the case of China, emission sources could be present in each indoor room rather than outdoor, and there may not be frequent air exchange from room to room.

Table 3. Relationship between indoor and outdoor concentrations of VOC and PAHs[a]

Compound	China		Japan	
	Summer	Winter	Summer	Winter
Benzene	-0.301	0.085	0.727**	0.517**
Toluene	0.261	0.459	0.825**	0.465*
4-ring PAHs	0.389	0.497*	0.425*	0.258
5-ring PAHs	0.784**	0.843**	0.895**	0.565**
6-ring PAHs	0.767**	0.896**	0.766**	0.500*

[a] *$p < 0.05$, **$p < 0.01$

For PAHs, significant correlations between the concentrations of 5- and 6-ring PAHs in indoor and outdoor air were observed in both two countries ($p < 0.05$) (Table 3). The correlations for those in China were more significant in winter than in summer, however, opposite case was observed in Shizuoka. The results strongly supported the above views that concentrations of 5 to 6-PAHs were predominated by outdoor sources, especially during the winter in China and summer in Japan. The average ratios of indoor/outdoor PAH concentrations are also given in Figure 2. In general, the variation trend of indoor/outdoor ratios was similar in the two cities. The ratios of 4-ring PAHs were higher than 1.0 in summer in residential air of China, which indicated that there were indoor emission sources for the 4-ring PAHs. However, the ratios of 5- and 6-ring PAHs were generally lower than 1.0, especially in winter, indicating that they mainly come from outdoor emission sources.

4. Exposure Risks

Although personal exposures to air pollutants are generally greater than the indoor concentrations, no indoor air quality problem should be ignored because we spend most of our daily life indoors. Furthermore, indoor air quality is strongly influenced by the outdoor environment as described above. Therefore, not only indoor microenvironments but also the outdoors were considered in assessing human health risks.

Here we report our investigation of cancer risks from inhalation in terms of unit risk values for carcinogenic VOCs (benzene and ethylbenzene) and the geometric mean (GM) concentrations at each monitoring site during the summer and winter campaigns calculated using

$$\mathrm{LCR}_{ij} = \sum_{ij} \mathrm{UR}_j \times \mathrm{C}_{ij}, \quad [1]$$

where LCR_{ij} is the estimated risk from pollutant j at study site i, C_{ij} is the measured concentration of pollutant j at study site i, and UR_j is the inhalation UR for pollutant j.

The unit risk values of benzene and ethylbenzene used here were 2.9×10^{-5} and 5.0×10^{-7} respectively (Cal EPA, 2002; Caldwell et al., 1998). Table 5 shows the GM and the calculated LCR of each carcinogenic VOC at living room. Among the VOCs, benzene posed a significantly greater risk than ethylbenzene did; the LCR of benzene was approximately 50 times that of ethylbenzene. Comparing the LCRs, the values ranged from 4.5×10^{-6} to 2.4×10^{-4} in China, and from 7.6×10^{-7} to 2.6×10^{-6} in Japan; that is, the LCRs in China were approximately 10 times those in Japan. The risks calculated for China were considerably higher than (more than 10 times) those in indoor monitoring in Hong Kong (Guo et al., 2004c), Baltimore City (Payne-Sturges et al., 2004) and in three urban communities in Minneapolis, United States (Sexton et al., 2004). Benzene has been estimated to account for more than 40% of the LCRs in indoor environments (Guo et al., 2004c). Therefore, the actual risks could be even greater because the values estimated in the current study are limited to only a few VOCs.

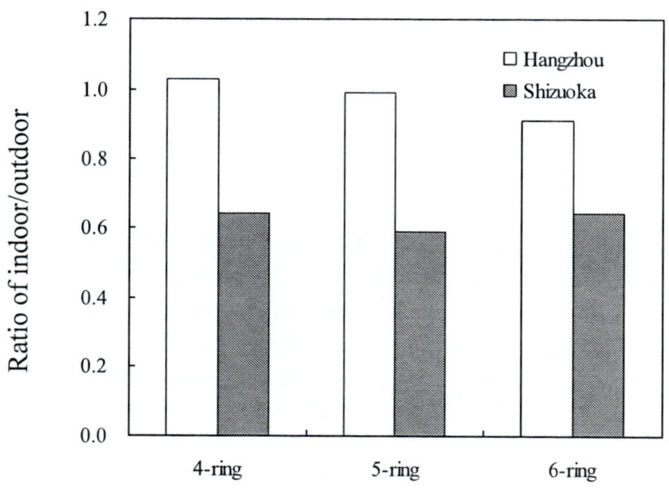

Figure 3. Average indoor/outdoor Ratios of PAH concentrations.

Table 4. LCRs of exposure to carcinogenic VOCs in indoor in China and Japan

	China		Japan		
	GM	LCA	GM	LCA	C/J
Benzene	8.18	2.4E-04	0.89	2.6E-05	9.2
Ethylbenzene	9.06	4.5E-06	1.52	7.6E-07	6.0

Table 5. Average TEF-adjusted concentrations of PAHs in residential air (ng/m^3)

PAHs	TEF	China		Japan	
		Summer	Winter	Summer	Winter
BaA	0.1	3.5	1.2	0.022	0.038
CHRY	0.01	0.073	0.061	0.0070	0.0086
BbF	0.1	0.14	0.56	0.12	0.14
BkF	0.1	0.069	0.24	0.047	0.044
BaP	1	0.61	4.4	0.65	0.85
DA	1	0.33	0.69	0.13	0.14
BP	0.01	0.012	0.043	0.0094	0.011
IN	0.1	0.065	0.49	0.09	0.096
∑TEF-PAHs		4.8	7.7	1.1	1.3

Of PAHs, toxicity equivalency factor (TEF) based on BaP could be used to assess the toxicity potencies of various PAHs in respect to inhalation cancer risks to humans (Nisbet and LaGoy, 1992). Average TEF-adjusted PAH concentrations (∑TEF-PAHs) in residential air are shown in Table 5. The results indicated that average ∑TEF-PAHs in Hangzhou were 4.5 times and 5.8 times higher than those in Shizuoka in summer and winter, respectively. In Hangzhou, average ∑TEF-PAHs were mainly contributed from BaA (72.9%) in summer and BaP (57.2%) in winter. However, average ∑TEF-PAHs were dominated by BaP (60.4% in summer and 64.0% in winter) in both two seasons in Shizuoka.

CONCLUSIONS

We have determined aromatic VOC and PAH concentrations in indoor microenvironments and outdoors in the summer and winter periods in two cities in Japan and China. The main objective of this study was to compare the occurrences of inspirable air pollutions for China and Japan, and reveal differences including those of pollution conditions and emission sources. Analysis of the indoor concentrations and seasonal variations of those pollutants indicated characteristic differences between the two countries.

- The concentrations of VOCs and PAHs in China were significantly higher than those in Japan in both summer and winter; the levels in China was extremely higher (>10-times) than that in Japan.
- Indoor VOCs in Japan were significantly affected by outdoor pollution, whereas the significant sources could be present indoors in China.
- Cancer risks estimated from the concentrations of carcinogenic VOCs and PAHs, the risks in China were approximately 10 times those in Japan.

Humans spend most of their daily life indoors, and thus indoor air quality is especially important in ensuring human health. To prevent health hazards caused by air pollution, urgent countermeasures for the reduction of potent indoor sources are required in China.

REFERENCES

Amagai, T., Ohura, T., Sugiyama, T., Fusaya, M., Matsushita, H., 2002. Gas chromatographic/mass spectrometric determination of benzene and its alkyl derivatives in indoor and outdoor air in Fuji, Japan. *J. AOAC Int.* 85, 203–211.

Boström, C.E., Alme′ n, J., Steen, B., Westerholm, R., 1994. Human exposure to urban air pollution. *Environ. Health Perspect.* 102, 39–47.

Cal EPA, 2002. *Toxicity Criteria Database.* Available at:. California Environmental Protection Agency, Sacramento, CA. *http://www.oehha.ca.gov/.*

Caldwell, J., Woodruff, T.J., Morello-Frosch, R., Axelrad, D.A., 1998. Application of health information to hazardous air pollutants modeled in EPA's cumulative project. *Toxicol. Ind. Health* 14, 429–454.

Caricchia, A.M., Chiavarini, S., Pezza, M., 1999. Polycyclic aromatic hydrocarbons in the urban atmospheric particulate matter in the city of Naples (Italy), *Atmos. Environ.* 33, 3731-3738.

Chan, C.-C., Ozkaynak, H., Spengler, J.D., Sheldon, L., 1991. Driver exposure to volatile organic compounds, CO, ozone and NO2 under different driving conditions. *Environ. Sci. Technol.* 25, 964–972.

Delfino, R.J., Gong, H., Linn, W.S., Hu, Y., Pellizzari, E.D., 2003. Respiratory symptoms and peak expiratory flow in children with asthma in relation to volatile organic compounds in exhaled breath and ambient air. *J. Expo. Anal. Environ.Epidemiol.* 13, 348–363.

Elbir, T., Cetin, B., Cetin, E., Bayram, A., Odabasi, M., 2007. Characterization of volatile organic compounds (VOCs) and their sources in the air of Izmir, Turkey. *Environ. Monit. Assess.* 133, 149–160.

Finlayson-Pitts BJ, Pitts Jr Jn. Indoor Air Pollution: Sources, Levels, Chemistry, and Fates. In *Chemistry of the Upper and Lower Atmosphere*, 2000, 844–870.

Guo, H., Lee, S.C., Chan, L.Y., Li, W.M., 2004a. Risk assessment of exposure to volatile organic compounds in different indoor environments. *Environ. Res.* 94, 57–66.

Guo, H., Wang, T., Louie, P.K.K., 2004b. Source apportionment of ambient nonmethane hydrocarbons in Hong Kong: application of a principal component analysis/absolute principal component scores (PCA/APCS) receptor model. *Environ. Pollut.* 129, 489–498.

Guo, H., Wang, T., Simpson, I.J., Blake, D.R., Yu, X.M., Kwok, Y.H., Li, Y.S., 2004c. Source contributions to ambient VOCs and CO at a rural site in eastern China. *Atmos. Environ.* 38, 4551–4560.

Hein, M.J., Waters, M.A., van Wijngaarden, E., Deddens, J.A., Stewart, P.A., 2008. Issues when modeling benzene, toluene, and xylene exposures using a literature database. *J. Occup. Environ. Hyg.* 5, 36–47.

Holian, A., 1996. Air toxics: biomarkers in environmental applications-overview and summary of recommendations. *Environ. Health Perspect.*104 (Suppl. 5), 851–855.

Jones, A.P., 1999. Indoor air quality and health. *Atmos. Environ.* 33, 4535–4564.

Kim, Y.M.,Harrad, S.,Harrison, R.M., 2001. Concentrations and sources ofVOCs in urban domestic and public microenvironments. *Environ. Sci. Technol.* 35, 997–1004.

Kume, K., Ohura, T., Noda, T., Amagai, T., Fusaya, M., 2007. Seasonal and spatial trends of suspended-particle associated polycyclic aromatic hydrocarbons in urban Shizuoka, Japan. *J. Hazard. Mater.* 144, 513–521.

Kume, K., Ohura, T., Amagai, T., Fusaya, M., 2008. Field monitoring of volatile organic compounds using passive air samplers in an industrial city in Japan. *Environ. Pollut.* 153, 649–657.

Li, S., Chen, S., Zhu, L., Chen, X., Yao, C., Shen, X., 2009. Concentrations and risk assessment of selected monoaromatic hydrocarbons in buses and bus stations of Hangzhou, China. Sci. *Total Environ.* 407, 2004–2011.

Liu, Y., Shao, M., Fu, L., Lu, S., Zeng, L., Tang, D., 2008. Source profiles of volatile organic compounds (VOCs) measured in China: part I. *Atmos. Environ.* 42, 6247–6260.

Loh, M.M., Houseman, E.A., Gray, G.M., Levy, J.I., Spengler, J.D., Bennett, D.H., 2006. Measured concentrations of VOCs in several non-residential microenvironments in the United States. *Environ. Sci. Technol.* 40, 6903–6911.

Low, L.K., Meeks, J.R., Mackerer, C.R., 1988. Health effects of the alkylbenzenes. I.Toluene. *Toxicol. Ind. Health* 4, 49–75.

Maliszewska-Kordybach, B., 1999. Sources, concentrations, fate and effects of polycyclic aromatic hydrocarbons (PAHs) in the environment. part A: PAHs in air, *Pol. J. Environ. Stud.* 8, 131-136.

Mastral, A.M., Callen, M.S., 2000. A review on polycyclic aromatic hydrocarbon (PAH) emissions from energy generation, *Environ. Sci. Technol.* 34, 3051-3057.

Na, K., Kim, Y.P., Moon, I., Moon, K.-C., 2004. Chemical composition of major VOC emission sources in the Seoul atmosphere. *Chemosphere* 55, 585–594.

Naumova, Y.Y., Eisenreich, S.J., Turpin, B.J., Weisel, C.P., Morandi, M.T., Colome, S.D., Totten, L.A., Stock, T.H., Winer, A.M., Alimokhtari, S., Kwon, J., Shendell, D., Jones, J., Maberti, S., Wall, S.J., 2002. Polycyclic aromatic hydrocarbons in the indoor and outdoor air of three cities in the US, *Environ. Sci. Technol.* 36, 2552-2559.

Nielsen, G.D., Larsen, S.T., Olsen, O., Lovik, M., Poulsen, L.K., Glue, C., Wolkoff, P., 2007. Do indoor chemicals promote development of airway allergy? *Indoor Air* 17, 236–255.

Nisbet, I.C.T. LaGoy, P.K. Toxic equivalency factors (TEFs) for polycyclic aromatic hydrocarbons (PAHs), *Regul. Toxicol. Pharm.* 16 (1992) 290-300.

Ohura, T., Noda, T., Amagai, T., Fusaya, M., 2005. Prediction of personal exposure to PM2.5 and carcinogenic polycyclic aromatic hydrocarbons by their concentrations in residential microenvironments. *Environ. Sci. Technol.* 39, 5592–5599.

Ohura, T., Amagai, T., Fusaya, M., Matsushita, H., 2004. Polycyclic aromatic hydrocarbons in indoor and outdoor environments and factors affecting their concentrations, *Environ. Sci. Technol.* 38, 77-83.

Ohura, T., Amagai, T., Senga, Y., Fusaya, M., 2006. Organic air pollutants inside and outside residences in Shimizu, Japan: levels, sources and risks. *Sci. Total Environ.* 366, 485–499.

Payne-Sturges, D.C., Burke, T.A., Breysse, P., Diener-West,M., Buckley, T.J., 2004. Personal exposuremeets risk assessment: a comparison ofmeasured and modeled exposures and risks in an urban community. *Environ. Health Perspect.* 112, 589–598.

Polzin, G.M., Kosa-Maines, R.E., Ashley, D.L., Watson, C.H., 2007. Analysis of volatile organic compounds in mainstream cigarette smoke. *Environ. Sci. Technol.* 41, 1297–1302.

Sexton, K., Adgate, J.L., Ramachandran, G., Pratt, G.C., Mongin, S.J., Stock, T.H., Morandi, M.T., 2004. Comparison of personal, indoor, and outdoor exposure to hazardous air pollutants in three urban communities. *Environ. Sci. Technol.* 38, 423–430.

Wang, Z., Bai, Z., Yu, H., Zhang, J., Zhu, T., 2004. Regulatory standards related to building energy conservation and indoor-air-quality during rapid urbanization in China. *Energy Build* 36, 1299–1308.

Zhang, K.H., Song, S., 2003. Rural–urbanmigration and urbanization in China: evidence from time-series and cross-section analyses. *China Econ.* Rev. 14, 386–400.

Zhu, L.Z., Wang J., 2003. Sources and patterns of polycyclic aromatic hydrocarbons pollution in kitchen air, China, *Chemosphere* 50, 611-618.

In: Pollution in China
Editor: Michael I. Chang

ISBN: 978-1-61122-022-3
©2011 Nova Science Publishers, Inc.

Chapter 6

OZONE POLLUTION IN CENTRAL-EAST CHINA

Wenpo Shan[1,2*] and Yongquan Yin[2]

[1]College of Chemistry and Environmental Science, Hebei University, Baoding 071002, PR China

[2]School of Environmental Science and Engineering, Shandong University, Jinan 250100, PR China

ABSTRACT

The rapid economic growth, urban expansion, and transportation facility have led to severe air pollution problems in central-eastern China. However, very limited studies of air pollutions in this region have been conducted. In this chapter, we will analyze the surface ozone pollution in central-eastern China based on the measurement results at three observational sites (Jinan, an inland city; Mt. Tai, a mountain site with the altitude of 1534m a.s.l.; Yantai, a coastal city) from April 2003 to April 2006. The main results and conclusions are: (1) ozone pollution was severe in this region, especially in summer; (2) the three typical sites have obviously different characteristics of ozone pollution with each other; (3) ozone pollution at the mountain site and coastal site were both more severe than that at the urban site, which associated with the ozone depression process of urban atmosphere; (4) surface ozone present higher levels in spring and summer than that in autumn and winter at the urban site; (5) besides temperature and solar radiation, sea-land breeze circulation is an important factor influencing the ozone level at the coastal site, and maritime wind often induce high ozone levels; (6) the diurnal variation magnitude of ozone concentrations at the mountain site was much smaller than the urban site due to the lower local pollutant emissions. The study in his chapter can contribute to a better understanding of the ozone pollution in the vast central-eastern China caused by the anthropogenic activity.

Keywords: Ozone pollution; Urban site; Mountain site; Coastal site; Central-east China

*Corresponding author. Tel: +86-312-5079422; Fax: +86-312-5079422. *E-mail address:* wpshan@yahoo.com.cn.

1. INTRODUCTION

Surface ozone is the most abundant photochemical oxidant in the troposphere and is often used as an indicator of photochemical pollution. Elevated surface ozone could cause damages to human health, vegetation, and materials (e.g., Burnett et al., 2004; McLaughlin and Downing, 1995; Wang et al., 2005). Besides, surface ozone can enhance the greenhouse effect due to the absorption of solar radiation in the 9-10ìm wavelengths (e.g., Beaney and Gough, 2002). Surface ozone and related precursors and meteorological factors have been monitored broadly, especially in North America and Europe (Solomon et al., 2000). These studies have comprehensively analyzed the processes and impacts of ozone pollution.

China is the most populous country in the world and is one of the top energy consuming countries. Rapid economic development, urban expansion, and transportation facility contributed increased consumption of fossil fuel in the past three decades in China, which resulted in increased emission of air pollutants, especially in the eastern coastal region. The studies of surface ozone in this region were mainly focused on the Yangtze Delta and Beijing-Tianjin Region. Elevated ozone levels have been observed at some non-urban sites in the Yangtze Delta (Shan et al., 2010; Wang et al., 2006; Vincent and Wang, 2001). We have also observed higher ozone level during maritime winds than during continental winds in each season, while the primary pollutant levels presented the opposite result, at a coastal site in the Yangtze Delta, nearby Shanghai (Shan et al., 2010). Ozone levels above 150ppbv were observed at a background site (Shangdianzi station) nearby Beijing (Meng et al., 2009). There are limited observations and studies of air pollutants in the vast area between Yangtze Delta and Beijing-Tianjin Region (Wang et al., 2007).

We have done some observational studies in central-east China from 2003 to 2006. The observational sites (Jinan, an inland city; Mt. Tai, a mountain site with the altitude of 1534m a.s.l.; Yantai, a coastal city) were all located in Shandong province, which is one of the highest energy consumption provinces in China. The observational results have been published in some papers (e.g., Shan et al., 2008; Shan et al., 2009a; Shan et al., 2009b; Yin et al., 2005; Yin et al., 2006a; Yin et al., 2006b; Yin et al., 2010). The content in this chapter is mainly based on the reanalysis of the observational data and syntheses of the previously published works.

2. STUDY SITE AND TECHNIQUES

The observational study was conducted at Jinan from April 2003 to April 2006, Mt. Tai from July to September 2003, and Yantai from July to September 2005. The locations of these observational sites were presented in Figure 1.

Jinan is the capital of Shandong Province, with an urban population of 2.7 million and an urban built-up area of 190km^2 in 2003. The Measurement at Jinan was carried out on the campus of Shandong University (36°42' N, 117°08' E, 34.5m a.s.l.), which is in the eastern area of Jinan, lying in the second ring road. More information about the observational site and Jinan city could be found in Shan et al, 2008. The measurement at Mt. Tai was carried out in Riguanfeng meteorological observation station (1534m a.s.l.), located at the summit of the mountain. More information about the observational site and Mt. Tai could be found in Gao et al., 2005 and Yin et al., 2006b. The measurement at Yantai was conducted in the Environmental Monitoring Station of Yantai Economic and Technological Development Zone, about 200m to the south of the Yellow Sea. More information about the observational site could be found in Yin et al., 2010.

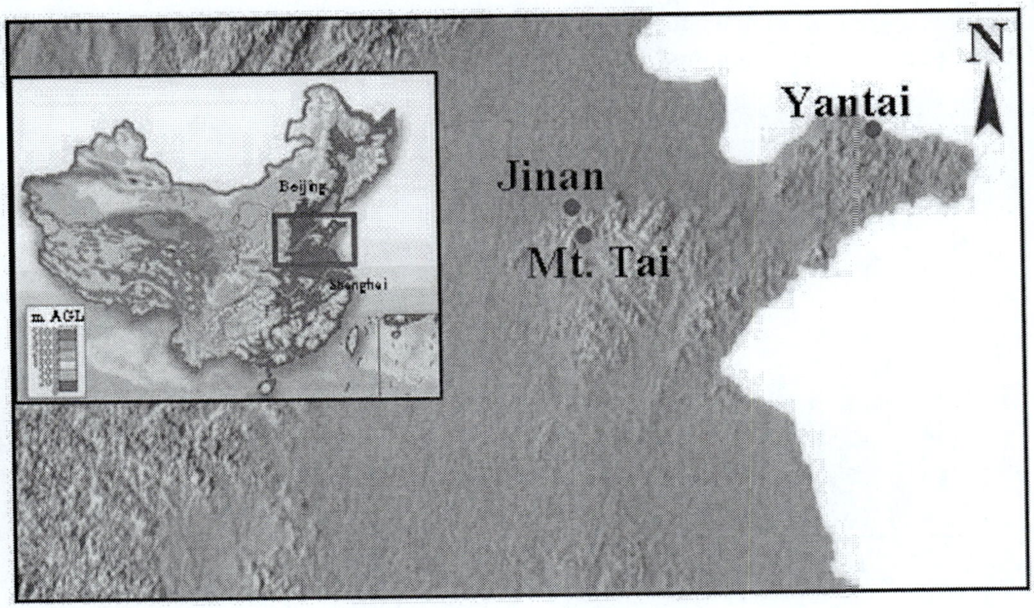

Figure 1. A map showing the locations of the observational sites (Jinan, Mt. Tai, and Yantai).

At the three sites, ozone was all continuously measured by a UV photometric analyzer (TEI, model 49C). Air samples were collected through teflon inlet tubes, with a particulate filter to prevent particles from entering into the instruments. Meteorological data used in this study were received from China Meteorological Data Sharing Service System (http://cdc.cma.gov.cn/) and local meteorological stations. An internet-based model of Hybrid Single-particle Lagrangian Integrated Trajectory (HYSPLIT, Version 4.8) was used to calculate back trajectories. The model was developed by the National Oceanic and Atmospheric Administration (NOAA) Air Resource Laboratory (http://www.arl.noaa.gov/ready/hysplit4.html).

3. Results and Discussion

3.1. Ozone Pollution at the Urban Site

3.1.1. Time Series of Ozone Concentrations

Figure 2 presents the time series of hourly averaged ozone concentrations from April 2003 to April 2006 at Jinan. Ozone concentrations exceeding the Chinese National Ambient Air Quality Standard Grade 2 (hourly average concentration 0.20 mg/m^3, about 100ppbv) were frequently observed in each year, with the maximum 1h value (147.83ppbv) observed during summer 2005. There is a clearly variation tendency of daily averaged ozone concentrations in the study period of a year, despite some fluctuations caused by changing weather conditions. Ozone presents maximum level in summer and minimum in winter. This variation pattern is almost the typical mode that observed at big cities in the Northern Hemisphere. However, the summer maximum of ozone in Jinan is just contrary to some sites under the impact of summer monsoon, such as the sites in Pearl River Delta, where are dominated by summer minimum (Zheng et al., 2009).

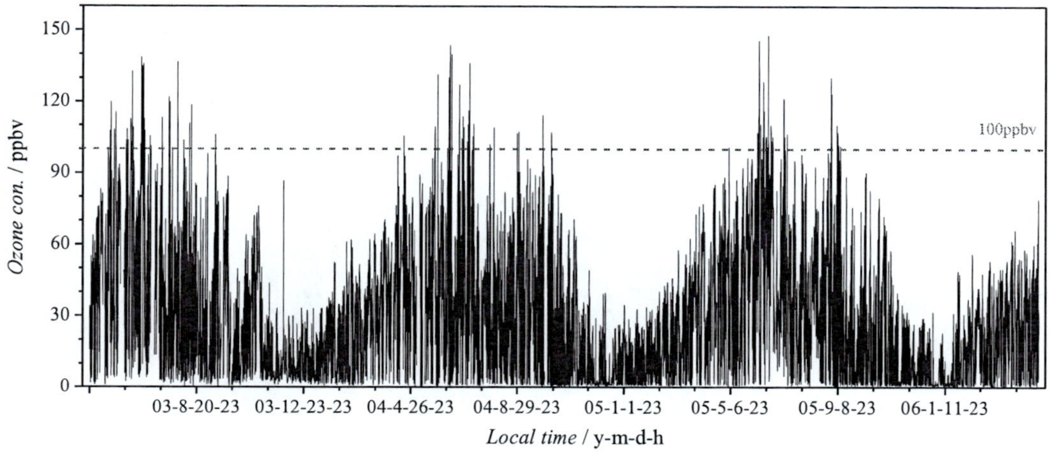

Figure 2. Time series of hourly averaged ozone concentrations from April 2003 to April 2006 at Jinan.

3.1.2. Seasonal and diurnal variations

Diurnal variations of ozone concentrations according to seasons averaged from winter 2003 to autumn 2005 are presented in Figure 3. Ozone diurnal variation of each season showed a similar pattern (a typical pattern for polluted urban areas characterized by high concentrations during mid or afternoon, low concentrations during late night or early morning, and big variation magnitude between daytime and nighttime), but the magnitudes of variations were different. Daytime hourly averaged ozone concentrations showed significant seasonal differences with a clear order of summer > spring > autumn > winter. However, ozone concentrations during nighttime from 22:00 to 6:00 in spring were higher than that in summer. Ozone spring maximum is a common phenomenon in the Northern Hemisphere. Although there are still some debates for the origins of this phenomenon, stratosphere-troposphere exchange and photochemistry are thought to be two main contributing processes (Monks, 2000).

3.1.3. Air masses classification and cluster analysis

The 96h backward trajectories of air masses arriving at the site during the ozone episode (with ozone concentration > 100ppbv) season of 2004 (from 16 April to 15 October) were calculated and clustered into 6 groups (see Figure 4), by considering their origins, paths, altitudes, and speeds of transport (Shan et al., 2009a).

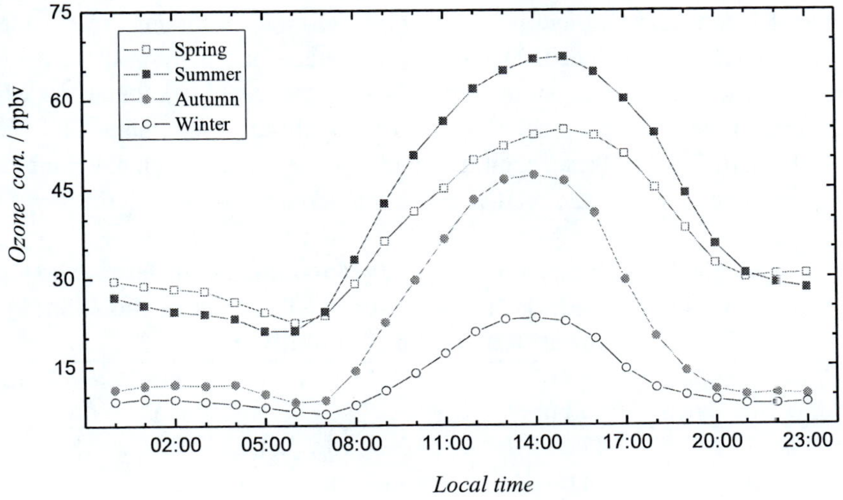

Figure 3. Diurnal variations of ozone concentrations at Jinan according to seasons averaged over the period from winter 2003 to autumn 2005.

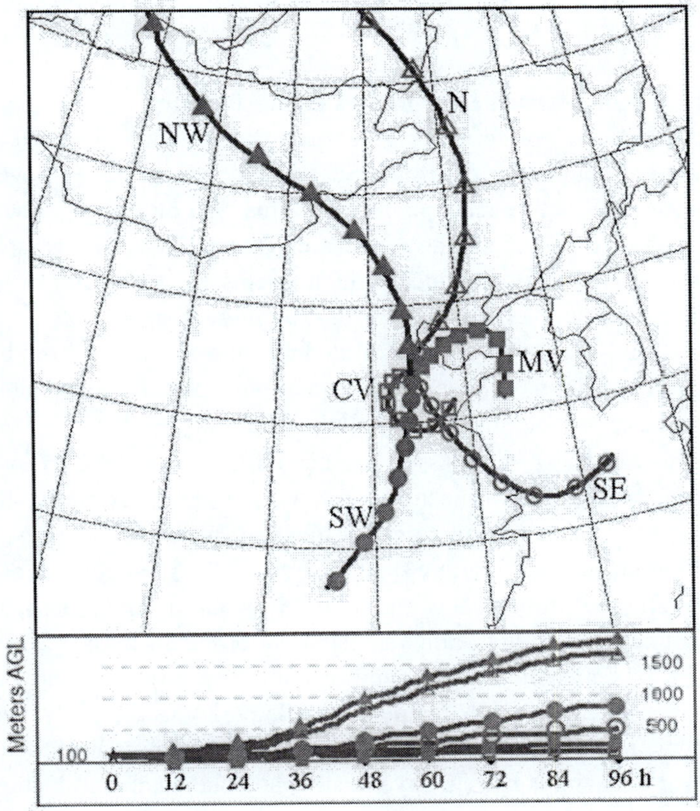

Figure 4. Six types of air masses arriving at Jinan during the episode season of 2C04, classified by 96h backward trajectories.

(1) NW—a cluster of continental air masses originates from the northwest areas, such as Central Asia, Siberia and Mongolia, with high altitudes, passing over northern or northwestern China with high speed. This cluster accounts for 9.8% of all the air masses in the examined days. However, none air mass of this group appeared in summer (i.e., from June to August), and none episode was related to this cluster (see table 1). The typical meteorological condition for these continental air masses is low relative humidity, little rainfall, long sunshine duration, and relatively low temperature.

Table 1. Meteorological conditions and ozone concentrations in different clusters. PRF: percentage of the days with rainfall; T: temperature; SD: sunshine duration; WS: wind speed; RH: relative humidity.

Cluster type	Total days[a]	Episode days	O_3[b] / ppbv	PRF /%	T[b] / °C	SD / h	WS /$m \cdot s^{-1}$	RH / %
NW	18	0	37.2 / 67.6	19.0	18.9 / 24.7	9.1	3.4	45.1
N	31	2	34.0 / 66.7	19.4	21.0 / 26.9	5.8	3.1	61.1
CV	67	14	39.1 / 81.1	22.1	23.6 / 29.2	6.2	2.7	62.4
MV	15	1	33.0 / 65.3	50.0	20.2 / 25.4	4.6	2.7	71.1
SW	17	2	38.7 / 66.2	41.2	27.1 / 31.5	5.8	3.9	66.4
SE	30	4	39.1 / 74.9	30.0	24.7 / 29.6	4.3	3.0	69.3

[a]Five days were not included due to data missing; [b]Daily averaged value / Daily maximum value.

(2) N—a cluster of continental air masses comes from the north or northeast areas with high altitude, passing over northern or northeastern China with high speed. Most of these air masses spend several hours over the Bohai Sea, and the others passing over Beijing-Tianjin region with low altitudes before reaching the site. This type of air masses was observed in 31 days, accounting for 16.9% of all the air masses (see table 1). There are two continuous episode days (1 and 2 July) associated with this cluster, which might be caused by the transport of pollutants from the polluted Beijing-Tianjin region.

Actually, NW and N are two clusters of cold air masses coming from high latitude areas with dry weather condition. Under these conditions, the transport pathways determine whether the air masses pick significant air pollutants or not.

(3) CV— this cluster of air masses has no defined path, located in low altitude and moving slowly or with a loop trajectory over the continental vicinity. It has the highest frequency in the six clusters, accounting for 36.6% of all the air masses, and the monthly differences in its frequency are relative small (see table 1). Most episode days are associated with this cluster, and the ozone concentration in this group is the highest. Ozone episodes in this group are unlikely due to long-range transport since the air masses keeping in the vicinal lower atmosphere. The meteorological conditions of the episode days show that, there was no rainfall, low winds (2.2m/s), high temperature (25.6/32.0°C for daily averaged/maximum value), and sufficient sunshine (7.2h) during these days, which indicate that episodes in this group might were mainly caused by local photochemical production and pollutant accumulation.

(4) MV— this cluster of air masses have no defined path, located in low latitude and having a short pathway or with a loop trajectory with most time spending over the marine

vicinity. There are fewer days attributed to this group than CV. Just like CV, low wind speed is a typical meteorological characteristic in MV since both groups are comprised by slowly moving air masses. On the other hand, different with the continental air masses in CV, MV is a group of marine air masses. Therefore, MV has higher relative humidity, more rainfall, shorter sunshine duration, and lower temperature, as well as fewer episode days (see table 1). There was only one episode attributed to this group, observed on 28 May, which was related to the long sunshine duration (11.1h) and low wind (1.8m/s).

(5) SW—air masses of this cluster come from southwest, mainly passing over the inland area of southern China. All of the air masses in this group presented before August, and most of them appeared in July (11 days). High wind speed and temperature are the meteorological characteristics of this group. Two episode days were observed in this group, both associated with slow moving air masses.

(6) SE—air masses of this cluster originate from the coastal area of eastern and southeastern China or East China Sea with different transport speeds. Four episode days were observed in this group. All the trajectories of air masses in these four days have passed over the highly polluted Yangtze Delta region. Except for the nighttime episode of 25 May, meteorological conditions in other three days were all favorable for photochemical production. There were also four low ozone days in this group with the daily maximum ozone concentrations lower than 40ppbv. All the trajectories in these low ozone days came from East China Sea and mainly passed over the marine region. Therefore, different pathways of the air masses could cause distinct pollution results.

3.2. Ozone Pollution at the Mountain Site and the Comparison with the Urban Site

3.2.1. Overall Characteristics

Table 2 shows some statistical characteristics of the hourly averaged ozone concentrations in each month from July to September 2003 at Mt. Tai. The mean concentrations of ozone in different months follow the order of July > August > September. The high ozone pollution during summer in this region is mainly associated with the favorable meteorological condition, such as intense solar radiation and high temperature. The high ozone level observed at Mt. Tai can be explained by the transport of pollutants to the summit of the mountain through mountain-valley breeze and/or the development of convective planetary boundary layer (Gao et al., 2005).

Table 2. Statistical Characteristics of hourly averaged ozone concentrations at Mt. Tai according to month from July to September 2003.

Season	Mean / ppbv	Minimum / ppbv	Maximum / ppbv	n
July	65.65	35.60	119.82	648
August	59.67	18.88	102.52	744
September	56.78	16.92	93.13	480

3.2.2. Day-to-day Variations

Figure 5 presented the day-today variations of the ozone concentrations from July to September 2003 at Mt. Tai, as well as that at Jinan for comparison. The daily averaged ozone levels at Mt. Tai were almost all higher than that at Jinan, which was related to the lack of ozone depression processes. The daily variation trends of ozone at the two sites were similar with each other, which is associated with the relative short distance between them (about 70km).

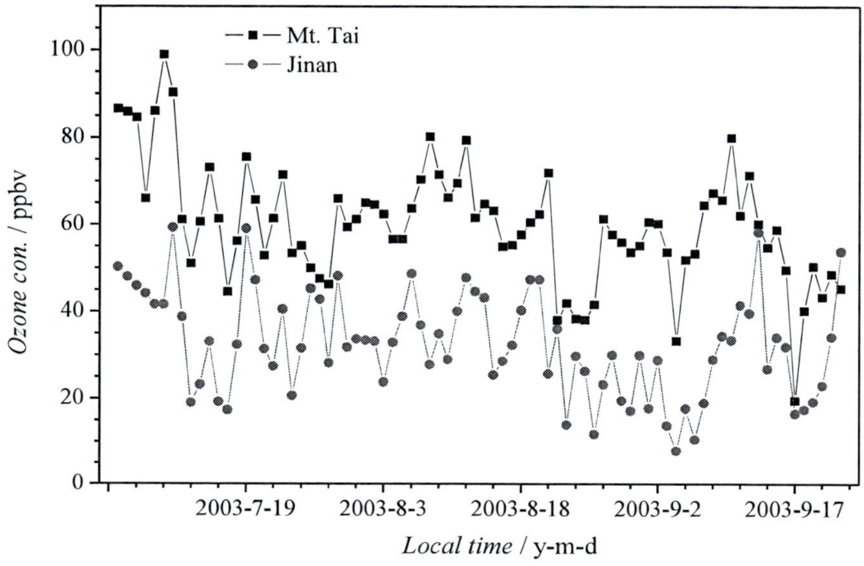

Figure 5. Variations of daily averaged ozone concentrations observed at Mt. Tai and Jinan from July to September 2003.

3.2.3. Diurnal Variations

Figure 6 presented the diurnal variations of the ozone concentrations averaged from July to September 2003 at Mt. Tai, as well as that at Jinan for comparison.

The diurnal variation of ozone at Mt. Tai presented the minimum level (56.18ppbv) at 10:00, while there was no obvious peak level. Different with Mt. Tai, the diurnal variation of ozone at Jinan started increasing rapidly coinciding with the increase of solar radiation from early morning to afternoon, reaching the peak value at 14:00, and then continuously decreased and kept relatively steady and low at night. This loss of ozone at night is attributed to in situ destruction of ozone by the well-known reaction between ozone and NO, and the surface deposition. The minimum value in the day appeared during early morning hours, near sunrise.

The magnitude of the diurnal variations was just 7.44ppbv at Mt. Tai, evidently lower than the magnitude of 46.68ppbv at Jinan, which can be explained by the different atmospheric environment and ozone pollution inducing process. The levels of primary pollutants, such as CO (see Figure 7), at Mt. Tai is clearly lower than that at Jinan, so the influences of ozone production and depression processes induced by atmospheric chemical reactions are weak at Mt. Tai. The ozone pollution at Mt. Tai is mainly attributed to the transport of photochemically processed boundary layer air masses (Gao et al., 2005), while the ozone pollution at Jinan is strongly influence by local photochemical reactions.

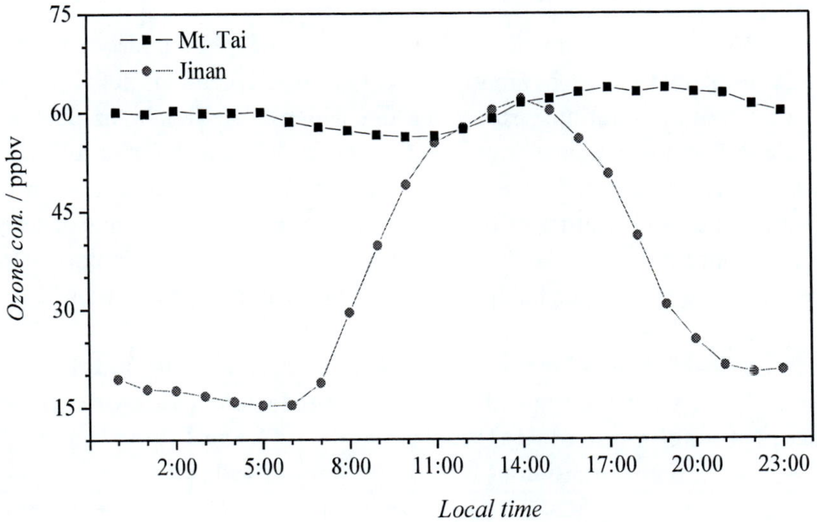

Figure 6. Diurnal variations of ozone concentrations at Mt. Tai and Jinan from July to September 2003.

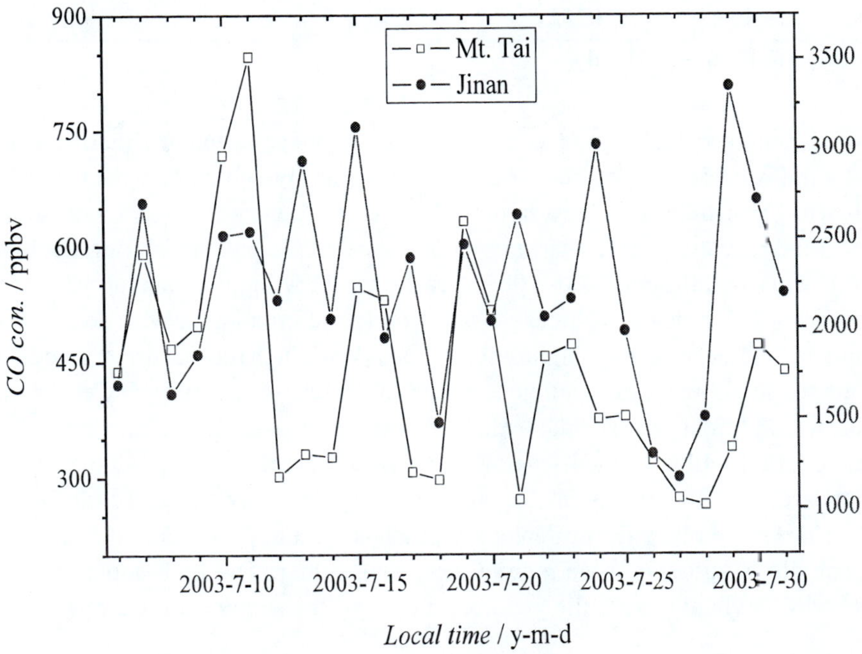

Figure 7. Variations of daily averaged ozone concentrations observed at Mt. Tai and Jinan during July 2003.

3.3. Ozone Pollution at the Coastal Site and the Comparison with the Urban Site

3.3.1. Overall Characteristics

Table 3 shows some statistical characteristics of the hourly averaged ozone concentrations at Yantai in each month from July to September 2005, as well as the percentages of the concentrations exceeding the Chinese National Ambient Air Quality Standard Grade 2. The mean concentrations of ozone in different months followed the order of September > July > August. However, both levels of temperature and solar radiation in different months followed the order of July > August > September, which is different with the order for ozone concentrations. This result indicates that some other important factors also contribute to the ozone level at Yantai besides temperature and solar radiation.

Table 3. Statistical results of measured ozone concentrations at Yantai from July to September 2005. PEC: Percentage of hourly averaged ozone concentrations exceeding the Chinese National Ambient Air Quality Standard Grade 2 (hourly average ozone concentration of 0.20 mg/m^3, about 100ppbv).

Month	Mean / ppbv	Median / ppbv	Maximum / ppbv	PEC / %	n	SD
July	41.02	38.11	150.98	14	622	25.15
August	39.64	39.27	115.28	11	719	21.44
September	41.43	38.24	135.81	22	679	25.44
Total	40.62	38.82	150.98	47	2020	24.00

[a]values for July are from 5 to 31 July.

According to the calculation, the average ozone concentration for maritime wind and continental wind were 50.52ppbv and 37.53ppbv respectively, which indicate that ozone level associated with maritime wind is higher than that associated with continental wind. According to the statistics, more maritime wind cases were observed during daytime (8:00-19:00) than during nighttime (20:00-7:00) in each month, while continental wind cases were contrary, indicating that sea-land breeze circulation (wind circulation with sea breeze during daytime and land breeze during nighttime in a day) might have occurred frequently. The frequency of sea-land breeze circulation occurred in August is obviously lower than the other two months, which might be associated with the observed August low ozone level.

The three months all presented several ozone values exceeding the National Ambient Air Quality Standard, with the maximum 1h value (150.98ppbv) observed at 17:00 July 26. Most exceeding values were observed in September, indicating that photochemical pollution at the observational site is still serious in autumn besides summer. The high summer and autumn ozone level will inevitably cause the yield reductions of the summer time crops in the coastal areas.

3.3.2. Day-to-day Variations

Figure 8 presents the variations of daily averaged ozone concentrations from July to September at Yantai, as well as at Jinan for comparison. Compared with July and early

August, the ozone level showed a little elevated from late August to September at Yantai, while ozone level kept decrease from late August to September at Jinan. The daily variation trends of ozone at the two sites were similar with each other, which is associated with the similar meteorological conditions due to the relative short distance between them.

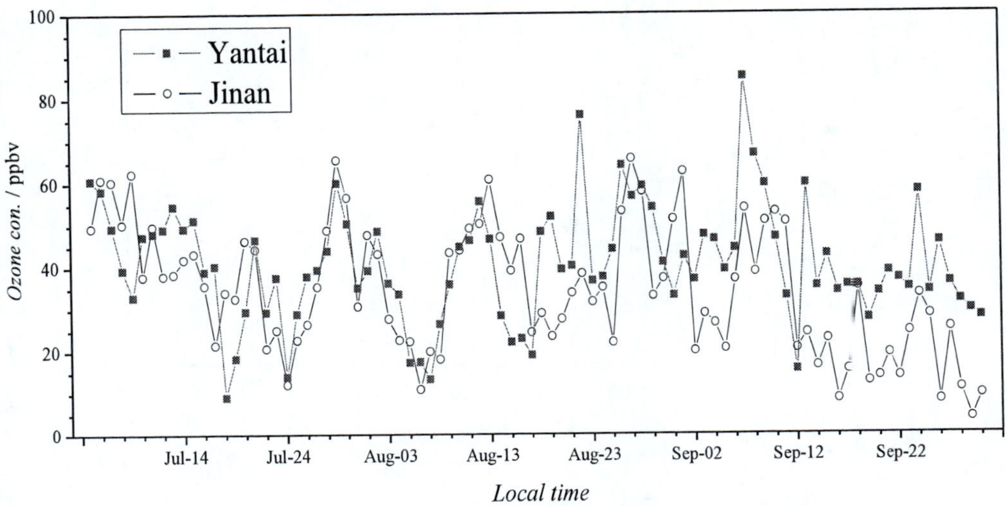

Figure 8. Variations of daily averaged ozone concentrations observed at Yantai and Jinan from July to September 2005.

3.2.3. Diurnal Variations

Figure 9 presents the averaged diurnal variations of ozone concentrations in each month (from July to September) at Yantai, as well as at Jinan for comparison. The diurnal variation of ozone at the two sites both showed a typical pattern for polluted areas. The two sites had similar diurnal ozone variation trends in July. However, the ozone levels were lower during daytime and higher during nighttime at Yantai than that at Jinan in August. In addition, the ozone levels were obviously higher both in daytime and nighttime at Yantai than that at Jinan in September. The different ozone levels in autumn indicate the different atmospheric environment between coastal site and inland site, even with a relatively short distance.

Compared with in July and September, the diurnal variation of ozone levels in August presents anomalously lower during daytime and higher during nighttime at Yantai (see Figure 9). Since the ozone level was higher during maritime wind than during continental wind, the anomalous diurnal variation of ozone in August might be related to the less sea-land breeze circulation. In addition, the highest ozone level of September might be associated with the highest frequency of maritime wind.

CONCLUSION

This chapter presents some observational studies in central-east China from 2003 to 2006. The observational sites (Jinan, an inland city; Mt. Tai, a mountain site with the altitude of 1534m a.s.l.; Yantai, a coastal city) were all located in Shandong province, which is one of the highest energy consumption provinces in China.

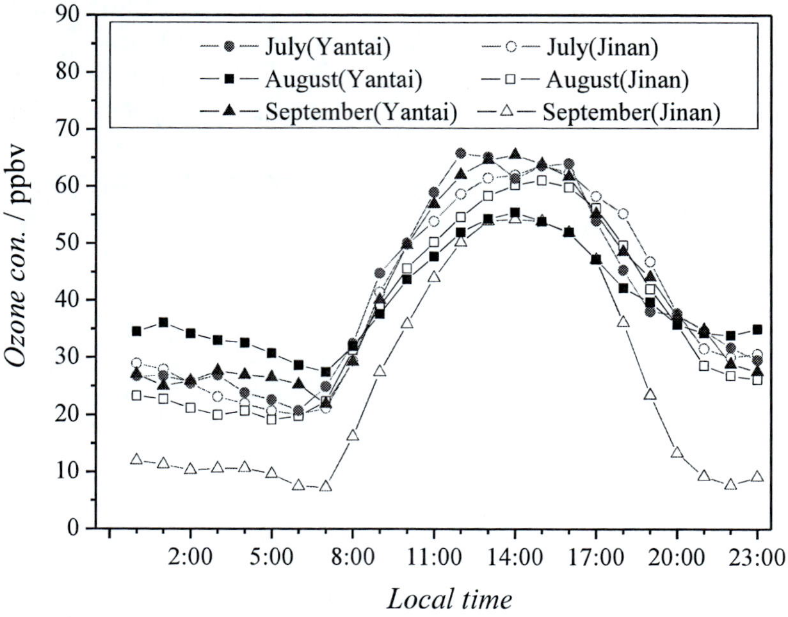

Figure 9. Diurnal variations of ozone concentrations in each month at Yantai and Jinan.

Surface ozone was continuously measured from April 2003 to April 2006 in the urban area of Jinan. Ozone concentrations exceeding the Chinese National Ambient Air Quality Standard were frequently observed in each year, with the maximum 1h value (147.83ppbv) observed during summer 2005. Ozone presents maximum level in summer and minimum in winter in a year. Ozone diurnal variation of each season showed a typical pattern for polluted urban areas characterized by high concentrations during mid or afternoon, low concentrations during late night or early morning, and big variation magnitude between daytime and nighttime. Cluster analysis of the 6 groups of backward trajectories shows that, most episodes were caused by local photochemical production and pollutant accumulation, and transport of pollutants from the highly polluted regions could significantly influence the air quality in the site, especially from Yangtze Delta region.

Surface ozone was continuously measured from July to September 2003 at Mt. Tai. The mean concentrations of ozone in different months follow the order of July > August > September. The daily averaged ozone levels at Mt. Tai were almost all higher than that at Jinan, which was related to the lack of ozone depression processes. The magnitude of the diurnal variations was just 7.44ppbv at Mt. Tai, evidently lower than the magnitude of 46.68ppbv at Jinan, which can be explained by the different atmospheric environment and ozone pollution inducing process.

Surface ozone was continuously measured from July to September 2005 at Yantai. The mean concentrations of ozone in different months followed the order of September > July > August. The ozone level was obviously higher at the coastal site than that at the inland site in September. Ozone pollution, with a maximum 1h concentration of 150.98ppbv, was severe during summer and autumn at the coastal site. Besides temperature and solar radiation, sea-land breeze circulation is an important factor influencing the ozone level at the coastal site, and maritime wind often induce high ozone levels.

This chapter revealed that the rapid developments of urbanization, industrialization and transportation have caused severe ozone pollution in central-eastern China. However, there are limited observations and studies of air pollutants in this region, between the highly polluted Yangtze Delta and Beijing-Tianjin Region. More detailed processes and impacts of the ozone pollution in this region need to be studied further.

REFERENCES

Beaney, G.; Gough, W.A. *Atmos. Environ.* 2002, *36*, 2319–2325.

Burnett, R.T.; Brook, J.R.; Yung, W.T.; Dales, R.E.; Krewski, D. *Environ. Res.* 1997, *72*, 24–31.

Gao, J.; Wang, T.; Ding, A.J.; Liu, C.B. *Atmos. Environ.* 2005, *39*, 4779–4791.

McLaughlin, S.B.; Downing, D.J. *Nature* 1995, *374*, 252–254.

Meng, Z.Y.; Xu, X.B.; Yan, P.; Ding G.A.; Tang, J.; Lin, W.L.; Xu, X.D.; Wang, S.F. *Atmos. Chem. Phys.* 2009, *9*, 927–936.

Monks, P. S. *Atmos. Environ.* 2000, *34*, 3545–3561.

Shan, W.P.; Yin, Y.Q.; Lu, H.X.; Liang, S.X. *Atmos. Res.* 2009a, *93*, 767–776.

Shan, W.P.; Yin, Y.Q.; Zhang, J.D.; Ding, Y.P. *Atmos. Res.* 2008, *89*, 252–261.

Shan, W.P.; Yin, Y.Q.; Zhang, J.D.; Ji, X.; Deng, X.Y. *Environ. Monit. Assess.* 2009b, *151*, 127–141.

Shan, W.P.; Zhang, J.D.; Huang, Z.X.; You, L.N. *Atmos. Res.* 2010, *97*, 26–34.

Solomon, P.; Cowling, E.; Hidy, G.; Furiness, C. *Atmos. Environ.* 2000, *34*, 1885–1920.

Vincent, T.F.; Wang, T. *Atmos. Environ.* 2001, *35*, 4947–4958.

Wang, H.X.; Kiang, C.S.; Tang, X.Y.; Zhou, X.J.; Chameides, W.L. *Atmos. Environ.* 2005, *39*, 3843–3850.

Wang, H.X.; Zhou, L.J.; Tang, X.Y. *J. Atmos. Chem.* 2006, *54*, 255–265.

Wang, X.K.; Manning, W.; Feng, Z.W.; Zhu, Y.G. *Environ. Pollut.* 2007, *147*, 394–400.

Yin, Y.Q.; Shan, W.P.; Ji, X.; Deng, X.Y. *B. Environ. Contam. Tox.* 2010, *85*, 10–14.

Yin, Y.Q.; Shan, W.P.; Ji, X.; Li, C.M.; Ma, G.X.; Cui, Z.J.; Wang T. *Environ. Pollut. Control* 2005, *27*, 716–718. (In Chinese, with an abstract in English).

Yin, Y.Q.; Shan, W.P.; Ji, X.; You, L.N.; Su Y.C. *Environ. Sci.* 2006a, *27*, 2299–2302. (In Chinese, with an abstract in English).

Yin, Y.Q.; Shan, W.P.; Wang, T.; Ji, X.; Li, C.M.; Cui, Z.J. *Environ. Sci.* 2006b, *27*, 9–13. (In Chinese, with an abstract in English).

Zheng, J.Y.; Zhong, L.J.; Wang, T.; Louie, P.K.K.; Li, Z.C. *Atmos. Environ.* 2010, *44*, 814–823.

In: Pollution in China
Editor: Michael I. Chang

ISBN: 978-1-61122-022-3
©2011 Nova Science Publishers, Inc.

Chapter 7

NONPOINT POLLUTION CONTROL FOR CROP PRODUCTION IN CHINA*

Zhao-liang Zhu[1], Bo Sun[1], Linzhang Yang[1], Linxiu Zhang[2], and David Norse[3]

[1] Institute of Soil Science, Academia Sinica, Nanjing 210008, P. R. China
[2] Center for Chinese Agricultural Policy, Institute of Geographical Sciences and Natural Resources Research, Chinese Academy of Sciences, Beijing 100101, P. R. China
[3] Department of Geography, University College London, 4 Taviton Street, London, WC1H 0BT, United Kingdom

China is facing the challenge of feeding its large and increasing population from a limited and decreasing area of cultivated land while achieving a clean and safe environment (Brown, 1994). After the onset of the green revolution in the 1950s, increasing inputs of synthetic fertilisers, organic manures, pesticides, and herbicides was an efficient tool to ensure the high yield in agriculture over the world. China now is the biggest user of synthetic fertilisers in the world. However this agro-chemical based intensive agriculture contributes substantially to the emission of greenhouse gases such as CH_4 and N_2O (Bouwman, 2001) and the entry of pollutants (nutrients, pesticide, heavy metals) into water bodies and soils. These pollutants cause adverse effects on environmental quality and public health, for example, ozone depletion in the upper atmosphere, the eutrophication in lakes and streams (Xing and Zhu, 2000), the pollution of soil and food.

With the development of the new concept of sustainable agriculture in the middle 1980s and of ideas about a double green revolution in the late 1990s (Conway, 1994), the target for global agriculture became development of production systems that satisfied food safety, economic and environment protection objectives. This led to the development of many new techniques and integrated resource management practices to mitigate the adverse effects of

* A version of this chapter was also published in Pollution Control: Management, Technology and Regulations, edited by Horatio R. Velasquez. It was submitted for appropriate modifications in an effort to encourage wider dissemination of research.

intensive farming on the environment. However, controlling the non-point pollution at the regional and continental scale is a complex problem. First we need to use advances in bio-physical research on nutrient cycling in agro-ecosystems to develop efficient techniques and policy measures to control the loss of nutrients (Yang and Sun, 2008). At the same time, we need to undertake socio-economic research to set up effective policy and institutional mechanisms to transfer the above techniques and management practices to farmers (Zhu et al., 2006). This socio-economic research highlights the necessity of drawing on the experience of other countries in Europe and Asia.

1. Status of Non-point Pollution From Crop Production in China

1.1. Non-Point Pollution from Synthetic Fertilizers

The China's consumption of synthetic fertilizers has been increasing year by year since the early 1960s to feed her huge population (Figure 1), however its yield increase stopped in the 1990s (Editor Committee of China Agricultural Yearbook, 2001-2002). China is now the largest producer and consumer of synthetic fertilizer in the world. Total fertilizer consumption reached 46.3 million tonne in 2004, that is, over one-third of world consumption. The national average annual application rate is about 211 kg N ha^{-1} cropland[1)], which is the fourth highest in the world after the Netherlands, South Korea and Japan. In some Provinces the average is greater than 400 kg N ha^{-1}, and in some counties over 1000 kg N ha^{-1} for the vegetable lands.

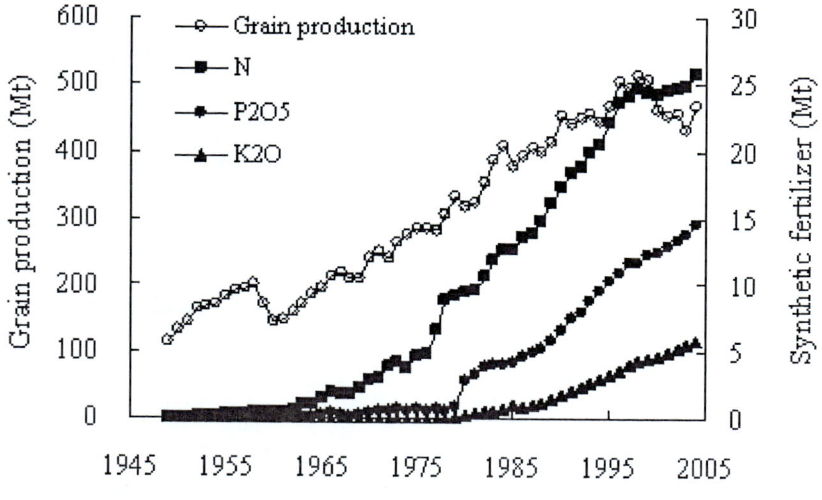

Figure 1. Grain production and synthetic fertilizer in China from 1949 to 2002.

1) The national average annual application rate was 167 kg N ha^{-1} in 2004 based on area under crop (total area of seeding).

Non-point pollution (Npp) from agriculture has become the dominant source of water pollution in China, and an important source of air pollution. Zhu, 2003 estimated that the total loss of nitrogen fertilizer from China's agriculture is about 19% (Table 1). The total input of chemical nitrogen (N) was 25.83 million tonne in China in 2004 and the total loss to the environment was some 4.93 million tonne, of which 1.29 million tonne entered the surface water, 0.52 million tonne passed down to the groundwater, and losses to the atmosphere were 0.28 million tonne (largely in the form of N_2O) and 2.84 million tonne as ammonia (NH_3).

These national averages hide considerable regional and cropping system variation in N losses to the environment (Tables 2). For example, leaching losses can be far greater in the high rainfall areas of southern China, and from irrigated intensive vegetable production. The research for urea showed that the average N leaching rate in North China was 2.1% and 2.7% for upland and paddy field respectively, while in South China was 8.2% and 5.1%.

Table 1. Fate of synthetic nitrogen in the agro-ecosystem during a crop season

Fate of nitrogen	Percentage	Environment impact
Runoff	5%	Surface water Eutrophication, Red tides
Leaching	2%	Nitrate in groundwater
Nitrification -denitrification	34% (of which 1.1% is N_2O-N)	Acid rain, Ozone Destruction, Global warming
NH_3 volatilization	11% (9% for Upland, 18% for rice paddy)	Atmospheric pollution, Acid rain
Crop recovery	35%	

The annual chemical nitrogen loss through leaching and runoff from farmland in China is about 1.73 million tonne (Zhu and Wen, 1994). The annual nitrogen input from agriculture to the Yangtse River and Yellow River is 92% and 88%, respectively, and about 50% of this comes from synthetic fertilizer. The annual soluble non-organic nitrogen (of which nitrate nitrogen NO_3-N accounts more than 80%) exported from Yangtse, Yellow and Zhu Rivers is now some 0.975×10^6 tonne (Duan and Zhang, 2000). Water bodies have been seriously polluted since the 1990's. About two-thirds of the water bodies in the seven river systems and three lakes (Taihu, Dianchi and Caohu) were in the worst quality class in 1999 (SEPA, 1999), and this state has not been improved in recent years (SEPA, 2004). The big lakes and all lakes within cities are in the middle quality class. Npp from crop production becomes a regional problem where the nitrate-N and phosphorous is carried by rivers to the sea, and causes eutrophication of coastal water. The estuaries and coastal water near cities can be seriously polluted and the frequency of red tides increased from 23 times in 1998 to 119 times in 2003 with a cumulative area of 14555 km^2 (SEPA, 1999, 2004).

Table 2. Estimated N output and N storage in the Yangtse, Huanghe and Zhujiang River valleys in 1995

Items	Amount (Tg N)		
	Yangtse	Huanghe	Zhujiang
N in the harvested crops	4.34	1.04	1.07
Denitrification in agricultural soils			
-Synthetic fertilizer N			
Rice fields	1.36-1.69	0.02-0.03	0.32-0.39
Uplands	0.41-0.92	0.22-0.50	0.07-0.10
-Organic fertilizer N	0.37-1.12	0.10-0.29	0.09-0.28
Subtotal	2.14-3.73	0.34-0.82	0.48-0.77
Storage in agricultural land			
-Synthetic fertilizer N			
Rice fields	0.82	0.012	0.19
Uplands	0.86	0.46	0.14
-Organic fertilizer N	1.83	0.48	0.83
Subtotal	3.51	0.95	1.16
N transported into water bodies			
-Anthropogenic and natural reactive N from input sources	2.87	0.65	0.62
-N from the excreta of people in cities and rural areas	0.92	0.18	0.22
Subtotal	3.79	0.83	0.84
NH_3 volatilization			
-From synthetic fertilizer N	1.32	0.17	0.29
-From excretive N of raised animals	1.0	0.28	0.27
Subtotal	2.32	0.45	0.56
Total	16-18	3.6-4.1	4.1-4.4

Xing and Zhu, 2002.

The nitrate content of groundwater and sources of drinking water has risen because of the increased applications of nitrogen fertilizer. The groundwater in 50% of China's cities suffers from this type of pollution, which is particularly serious in the north of China. For example, 38% of the drinking water wells in 16 counties in Jiangsu Province, Zhejiang Province and Shanghai City exceed the standard China has set for the nitrate-N content in drinking water ($\leq$ 20mg L^{-1}) (Zhang, 1999). About 58% of these wells are over the standard for nitrite-N content ($\leq$0.02 mg L^{-1}). In Beijing, Tianjin and Tangshan 50% of the sampling sites were over 11.3 mg L^{-1} (standard for Europe Union) in nitrate-N content - the highest reached 68 mg L^{-1} (Zhang, 1995) . In North-West China between Suide and Yulin 22% of the wells sampled (most of them for both drinking water and irrigation) exceeded the nitrate-N content standard). Thirty per cent of the 74 wells in the 24 counties in Guanzhong irrigation area and the drought plateau in the north of Wei River are over the standard (Lu et al., 1998; Jiang et al., 2002). It is clear that nitrate pollution of groundwater and drinking water may be a threat to people's health throughout China.

Npp and particularly nitrate pollution of groundwater is commonly very serious in the intensive vegetable growing areas that are found in or close to the suburbs of most towns and cities in China. A survey by the Chinese Academy of Agricultural Science at 800 sites in 20 counties of 5 northern provinces concluded that groundwater at 45% of the sites contained over 11.3 mg NO_3 L L^{-1}, 20% of the sites had over 20 mg NO_3-N L L^{-1}, and some had over 70 mg NO_3-N L^{-1}.

Projections suggest that the future nitrogen surplus[1)] from crop production will increase from about 154 kg ha^{-1} in 2004 to 179 kg ha^{-1} in 2015 (Figure 2), and hence the risk of non-point pollution will increase (Shen et al., 2005; Sun et al., 2008). Using estimates of the average nitrogen surplus in 2001, the Task Force concludes that the high risk area for fertilizer application is mainly in the coastal provinces of the southeast region and Hubei province in middle region. However, 15 provinces in the middle and southeast region in China, except Jiangxi and Shanxi province, will face high risks in 2015 if current policies and trends continue (Figure 3).

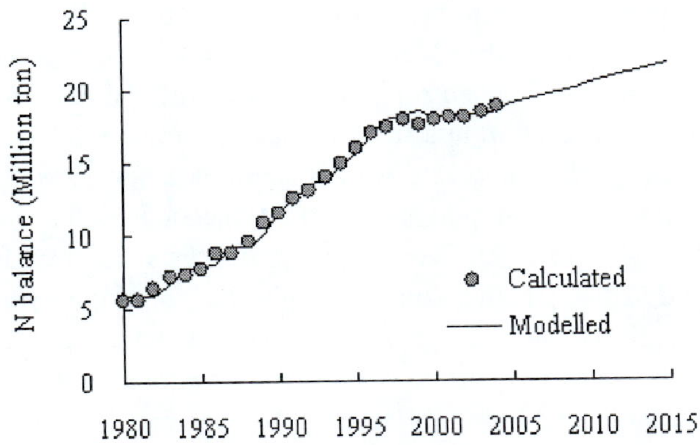

Figure 2. Prediction of nitrogen surplus in China's agro-ecosystem.

1.2. Status of Eutrophication of Chinese Lakes

China has made major advances in economic development, population growth and lake resource utilization over the past 25 years. However the prevention of lake pollution has lagged behind. Water environmental pollution, especially eutrophication, has become a serious problem. China has 4880 lakes, covering a total area of 83400km^2 and accounting for 0.8% of the country. About 50% of all of the lakes investigated in 2000 were eutrophic (Yuan, 2000), and for 75% of these lakes the eutrophication is getting worse.

1) The surplus is the difference between synthetic fertilizer input, biological N fixation, N from crop residues and the crop uptake, without taking losses through gas, runoff and leaching into consideration.

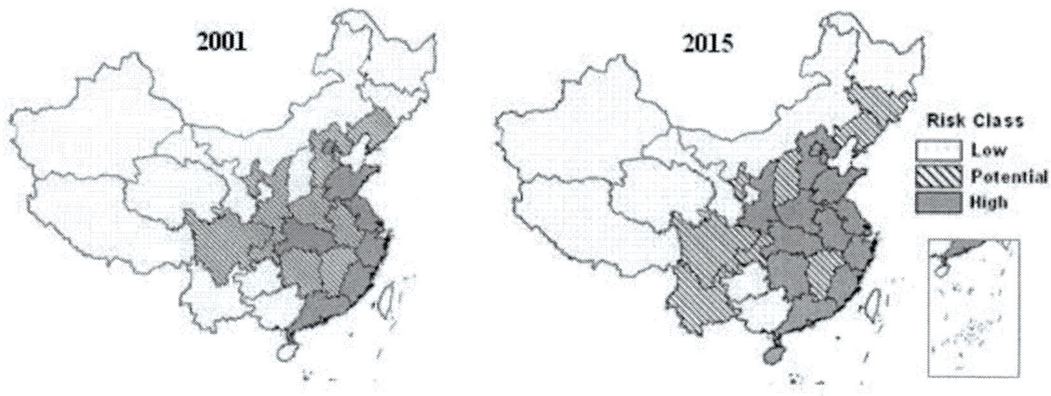

Figure 3. Evaluation of the risk of nitrogen application in 2001 and 2015.

According to preliminary statistics on the environmental capacity of 35 major Chinese lakes, about 5.65×10^6 tonnes waste water enter these lakes each day, accounting for 6.6% of all discharged over the whole country. Following the introduction of tighter regulations on and investment in the control of point source pollution, Npp is becoming the main factor increasing the eutrophication of Chinese lakes. In the case of three of China's major lakes with serious eutrophication, i.e. the Taihu, Dianchi and Chaohu Lakes, the total nitrogen (TN) in inputs from the non-point sources (including human and livestock excrements and domestic waste water) accounted for 59%, 33% and 63% of the whole lake loading, and the total phosphorus (TP) accounted for 30%, 41% and 73%, respectively (Li et al., 2001). In 1995, the non-point sources accounted for 55% of the total nitrogen loading and 28% of the total phosphorus loading without including the contribution from precipitation. The pollution from livestock/poultry production and aquaculture was about equal to that from industrial sources. Among the total pollution entering into the Tai-lake from agriculture source, the excrements from human and livestock accounted for 63%, atmospheric dry and wet deposition 28%, crop fertilization through runoff and leaching 9%, respectively (Zhu and Sun, 2008).

2. Reasons for Non-point Pollution from Crop Production in China

2.1. The Pressure for High Levels of Food Self-Sufficiency in China

China is a major agricultural and developing country with a population of 1.3 billion. It has 22% of the world's population but only 7% of the cultivated land of the world. The food production has increased substantially in the past 50 years, and this is largely because of progress in science and technology and institutional reform. Statistics show that the grain production per capita in China in 1961 was only about 60% of the world average, but by 1998 the productions of grain, meat and egg per capita were all above the world average with milk production being the main exception. Much of the increase in grain production was the result of greater use of synthetic fertilizers, and there is a significant correlation between the annual fertilizer application and the grain production. Consequently, non-point pollution from crop production also increased. If China is to satisfy the increasing demand for food in the future and it's objective of achieving a Well-Off-Society, China will need to keep the use of synthetic fertilizers and other production inputs at a high level, and therefore the pressure to the environment will increase unless there are effective measures to control Npp. The same requirements apply to pesticides.

China's entry to the WTO will increase pressures on the environment from agriculture while helping Chinese agriculture to integrate more with world markets. Research shows that the application rates of agriculture chemicals will increase unless management and extension systems improve. Free trade will affect the price of agricultural inputs and agricultural products, and consequently affect cropping patterns and the input levels of synthetic fertilizers and pesticides. The relatively lower price of some agriculture chemicals on the international market could induce farmers to apply more fertilizers and pesticides. Moreover, the high market value of vegetables and fruits will promote further the over-fertilization, and make it increasingly difficult to control Npp.

2.2. The Fast Development of Vegetable Production

Vegetable production has become one of the most profitable and fast growing agricultural sectors in China. It uses 80% less land than cereal crops yet in many provinces it provides over half of the total profits of crop production.

In 2002 the total area planted with vegetables in China was about 20 million ha, and it produced over 600 million tonne of vegetables. There are now over 150 counties in China with more than 20,000 ha of vegetables. However, this very successful expansion of vegetable production has caused very serious Npp. Excessive water and fertilizer inputs are very common. Since the mid 1990's there have been more and more reports of serious pollution problems due to excessive water and nutrient inputs, especially in Liaoning, Shandong, Heilongjiang, Jiangsu, Shanxi, Hebei, Beijing, Hunan, Tianjin etc.

The most important source of Npp resulting from vegetable production is the high inputs of irrigation water and fertilizer, especially nitrogen fertilizer (Ma et aL, 2000). In all Chinese provinces except Inner Mongolia the average synthetic fertilizer application rate commonly exceeds 200 kgN ha^{-1}, with the highest rate of 740 kg ha^{-1} occurring in Shangdong province (Figure 4). Surveys of the open vegetable fields in the Beijing area showed that the average N and P inputs from inorganic and organic sources was c.680 kg N ha^{-1} and c.440 kg P_2O_5 ha^{-1}, respectively. Even higher inputs of N were found in protected fields (greenhouses and polythene tunnels), i.e. 1380 kg N ha^{-1} input. Furthermore, these inputs do not take account of the N received from atmospheric deposition and in irrigation water, which can total as much as 200 kg ha^{-1}. The monitoring showed the annual wet deposition of N was 23.6 kg ha^{-1} in Shangdong and Hebei (Zhang et al., 2006). Nitrogen use efficiency was never higher than 10% and resulted in poor economic performance as well as Npp. In Shouguang county, Shangdong province the total nutrient input for cucumber and tomato was 2060 kg ha^{-1} N, 2530 kg ha^{-1} P_2O_5, 1590 kg ha^{-1} K_2O, respectively, with about half of the N, P and K input coming from organic manure. The total nutrient input rate was about 2-6 times actual crop demand (Figure 5). It is estimated that in 1997 the amount of synthetic fertilizer wasted by overuse in Shandong Province alone was about 118,000 tonnes N, 152,000 t P_2O_5, and 65,000 tonnes K_2O. These excess inputs lead to the eutrophication discussed in section I above.

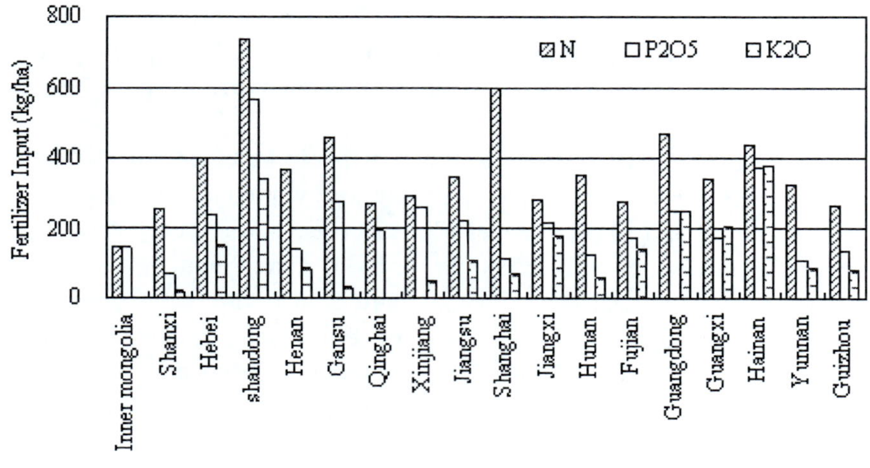

Figure 4. The total amount of synthetic fertilizer input in vegetable land in different provinces.

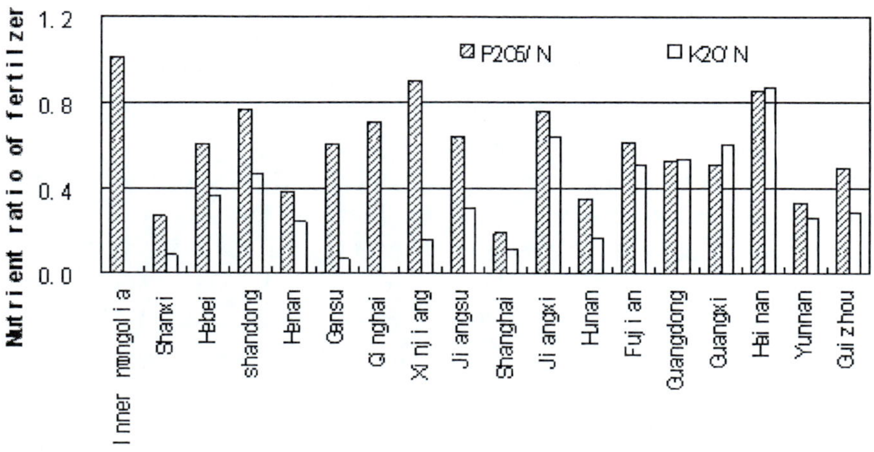

Figure 5. The nutrient ratio in the synthetic fertilizer input in the vegetable land in different provinces.

There are four main environmental impacts arising from intensive vegetable production, and are particularly severe with greenhouse vegetable production. First, there is the accumulation of nitrate in groundwater. The average nitrate content of the 0-4 metre soil profile below greenhouses in Beijing was 1230 kg N ha^{-1}, and the leached nitrate in groundwater could be over 200 kg N ha^{-1} each year. More than 90% of the shallow wells (<15m) surveyed in a greenhouse areas had nitrate levels above the maximum permissible concentration recommended by the WHO for drinking water. Secondly, high inputs of N fertilizer promote the gaseous loss of N. Experiments on vegetable land in Shouguang county Shangdong province showed that the loss of nitrous oxide (N_2O) during the spring season rose from c. 4.4 kg N ha^{-1} without synthetic fertilizer N inputs to 8.2 kg N ha^{-1} with a 870 kg N ha^{-1} synthetic fertilizer N input.

The third environmental problem arises from the high incidence of pests and diseases in vegetable production. Excessive N inputs are often the main reason for this high incidence, and in turn, this commonly leads to farms using too much pesticide, resulting in high pesticide residues on vegetables and in the environment. Dangerously high nitrate and pesticide residues are a constraint to the development of higher priced organic foods.

The fourth environmental problem is the damage that excessive inputs of synthetic fertilizers and irrigation water can cause to soil structure and soil quality. They can cause both biological and physico-chemical damage to soils, leading to acidification, secondary salinization and reduction of microbial activity. This damage lowers crop yields and may lead to farmers applying even more fertilizers to try to compensate for the reduced soil productivity, and thereby intensify Npp and the cycle of environmental degradation.

2.3. Unbalanced Nutrient Inputs to China's Agrosystems

The important role of synthetic fertilizer in China's agricultural production led to a very rapid climb in fertilizer consumption after the economic reforms of 1978. The average total fertilizer consumption during 1978-1980 was 11 million tonnes. It increased to 22 million tonnes in the late 1980s and to 40 million tonnes by the late 1990s. Total fertilizer consumption in China increased four fold during the first 20 years of economic reform. Consequently, in 1986 China overtook the United States as the world's largest consumer of fertilizers. China's farmers used 41 million tonnes of fertilizer in 2002, which is more than 25 percent of total world fertilizer use, even though China has less than 10% of the world's the total amount of area of arable land.

The ratio of organic fertilizer to synthetic fertilizer has decreased greatly. Prior to 1970 most of the nutrient inputs to farmland came from organic fertilizer. But the use farmhouse and domestic manure has decreased rapidly with economic development, the increase in off-farm employment and the rising cost of labour. These changes occurred first with nitrogen fertilizer, then in the 1980s with phosphate fertilizer, though most of the potassium still comes mainly from organic manure (Figure 6).

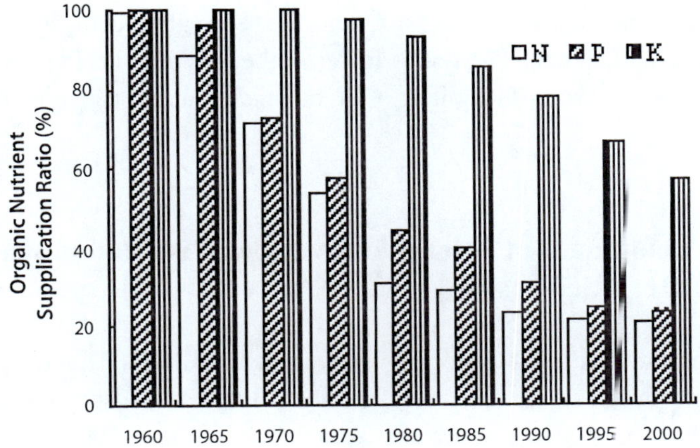

Figure 6. Nutrient inputs from organic fertilizers to agro-ecosystems from 1991 to 2000.

The nutrient ratio of mixed synthetic fertilizers is not in balance with crop and soil requirements (Yang and Sun, 2008). Although the ratio of N: P_2O_5: K_2O increased from 1:0.45:0.13 in 1991 to 1:0.52:0.20 in 2001, the proportion of potassium fertilizer should increase further. In 1999 only 1% of the counties had the correct nutrient ratio. Nitrogen ratios have been too high in most of regions in China since the 1970s, especially in eastern areas. Phosphorus ratios have changed from a deficit to small surplus (with a large surplus in some vegetable areas), but potassium is generally still in deficit. The total amount of nitrogen fertilizer increased rapidly before 1998, but is now nearly stable. It is forecasted to increase slowly and reach 33.4 million tonne in 2015 in China.

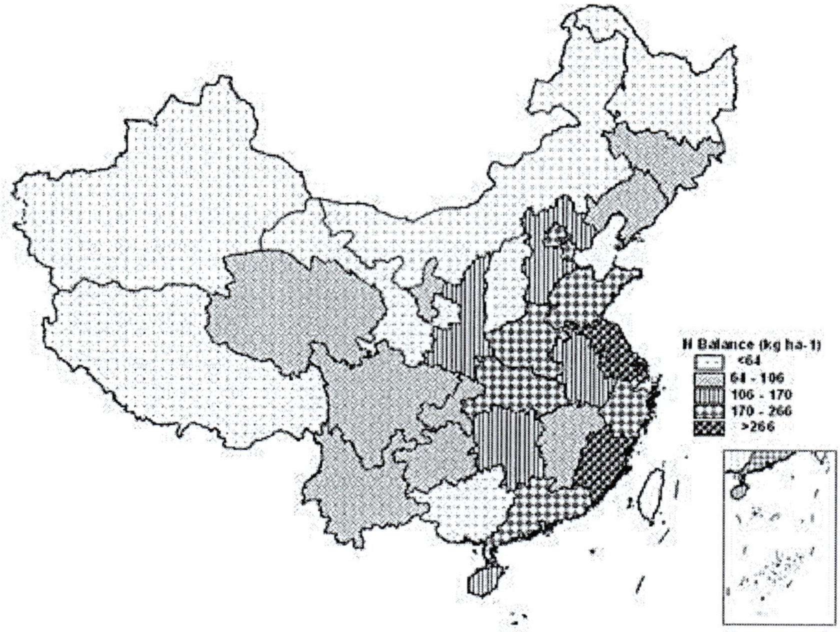

Figure 7. Average N surplus for China's provinces in 2002 to 2004.

Based on calculations of the nitrogen surpluses in the last 3 years (2002-2004), the highest risk to the environment from excess synthetic fertilizer N input is in the developed provinces (or metropolises) of southeast China, i.e. Shanghai, Jiangsu, Guangdong, Fujian, Beijing, Shandong and Henan. The risk is lower in the west and north provinces of China, i.e. Heilongjiang, Inner Mongolia, Xingjiang, Qinghai and Guangxi (Shen et al., 2005; Sun et al., 2008) (Figure 7).

2.4. Rapid Development of Intensive Livestock Production with Limited Treatment of Organic Wastes

Livestock production has developed very rapidly in China (Dong, 1998). Chicken, pig, dairy and beef production increased 5.6, 2.3, and 13.6 times, respectively during the period 1985-2002 (Table 3). Waste production from livestock was 690, 1420, 2700, and 4100 million tonnes in 1980, 1990, 2000 and 2002 respectively, and is predicted to reach to 6000 million tonnes in 2015. The total amount of wastes from livestock production in 2002 was over four times greater than production of organic pollutants from industry.

Table 3. Changes of livestock breeding in China (Million heads)

Animal	1985	1990	1995	2002	2015
Chicken	1978.9		4108.6	4735.2	13026
Pig	238.8	309.9	485.3	566.8	781.4
Dairy cow	1.6	2.7	4.2	6.9	23.7
Beef cattle	4.6	10.9	31.1	44.0	66.8

The production and use of organic fertilizer in China has received little attention in recent years. Huge amounts of manure, especially human wastes have become a source of pollutants rather than a resource to be recycled for fertilizer production. The utilization ratio by agriculture of manure nutrients depend on its collection ratio and the loss ratio during storage and transportation. In recent years, most of the human waste in towns and cities has been discharged directly into surface water bodies without any treatment, and rarely utilized for fertilizer. In rural areas the collection and reuse as fertilizer is better than in urban areas but lower than in the past. The greatest problem is with large scale livestock enterprises (especially pigs and cows) using concentrate feeds that discharge their wastes directly into water bodies. However, problems can also arise where there is no limit to or guidance on the application of manure to farmland, since overuse or badly timed use can damage soil ecological processes, and change soil from a pollutant filter to a source of pollutants.

Poor implementation of existing planning requirements or the lack of legal requirements for environmental impact assessments prior to the establishment or expansion of intensive livestock enterprises, and the lack of nationwide standards for waste discharges have led to the present situation where 90% of animal farms in China are equipped with no or inadequate waste disposal or treatment facilities. Moreover, with the lack of integration between livestock and crop production there is no waste management system to promote recycling, especially in peri-urban areas where the largest livestock enterprises tend to develop. Consequently, the animal wastes are directly discharged into the environment as waste instead of being used as a resource to be processed into organic fertilizers. The recycling ratios of the wastes of beef cattle, pigs, chickens, and dairy cows are 44%, 43%, 10% and 3%, respectively (Table 4).

In 1998 the total nitrogen and phosphorus content of various organic manures in China were over 16 and 7 million tonnes respectively. But the losses to environment were estimated to be nearly 11 million tonne of nitrogen and 2 million tonnes of phosphorus. Most of these losses were in the form of wastes directly discharged into surface water bodies, or as gaseous ammonia which can emit in substantial amounts. These losses were higher than those from synthetic fertilizer (1.24 million tonne for nitrogen) and hence were the main source of water pollution. This conclusion has been confirmed by isotope studies of rivers and lakes in the Tai lake region of Jingsu province (Xin et al., 2001).

Table 4. The production (recycling) ratios of manure from the wastes in 2002 and 2015

Animal	Manure production ratios (% of the total waste)		% N produced by all animal manure		% P produced by all animal manure	
	2002	2015	2002	2015	2002	2015
Beef cattle	44	48	47.6	46.7	26.4	27.3
Hog	43	34	26.3	31.5	33.3	34
Poultry	10	11	19.2	18.7	35.2	36.4
Dairy cow	3	7	6.8	3	5.2	2.4

With the booming of stockbreeding, the production of poultry manure was increasing continually and reach 2750 million tons in China in 2002. Sichuan Province had the highest production of poultry manure, which was followed by Henan and Shandong Provinces. The average load of poultry manure was 4.19 t ha^{-1} (based on total cultivated land area) in China. The higher environmental risk from poultry manure to the cultivated land was in Shanghai, Henan, Tianjin and Shangdong which had a load larger than 18 t ha^{-1}, the middle risk was in Beijng, Jiangsu, Hebei, Anhui and Hunan which had a load between 5 to 18 t ha^{-1}, and the other provinces had a low risk. The total N and P in the poultry manure were 15.3 and 6.4 million ton respectively. The amount of TN, TP , BOD and COD released to water body were 0.87 , 0.345, 6.0 and 6.74 million tons per year, respectively. Sichuan, Henan and Shangdong had a higher release of total N and total P (see Table 5) (Gao et al., 2006).

Table 5. The total N and P release from poultry manure in China in 2002

Total release (ton)		Province (Municipality)
N	>60000	Sichuan, Henan, Shandong
	60000-40000	Guangdong, Jiangxi, Hebei
	40000-20000	Anhui, Beijing, Guangxi, Guizhou, Heilongjiang, Hubei, Hunan, Inner Mongolia, Jiangxi, Jilin, Liaoning, Shanghai, Tianjin, Xinjiang, Yunnan
	<20000	Tibet, Qinghai, Gansu, Ningxia, Shanxi, Shaanxi, Chongqing, Zhejiang, Fujian, Hainan
P	>30000	Sichuan
	30000-20000	Henan, Hebei, Shandong
	20000-10000	Anhui, Beijing, Chongqing, Fujian, Gansu, Guangdong, Guangxi, Guizhou, Heilongjiang, Hubei, Hunan, Inner Mongolia, Jiangsu, Jiangxi, Jilin, Liaoning, Shaanxi, Shanghai, Tianjin, Xinjiang, Yunnan, Zhejiang
	<10000	Tibet, Qinghai, Ningxia, Shanxi, Hainan
No data		Hong Kong, Macao, Taiwan

2.5. Inadequate Agricultural Extension System

2.5.1. Under Investment

Compared with many other countries China's extension investment intensity in 1999 (ratio of agricultural extension as a percentage of total agricultural GDP) was only 0.49. This was only slightly higher than the average investment intensity of low-income countries in mid 1980s. It was much lower than the industrial nations (0.62 in 1980) and the USA (0.74 in 1990). More than 90% of the extension investment comes from local rather than central government, which will constrain extension services in poor areas where their activities are needed the most.

2.5.2. Misallocation of Investment Funds

Studies show that most of the funds allocated to extension services are used to pay staff salaries (80%). Moreover, very little of extension project money reaches local extension stations, because a large part of it tends to be retained by local government for other uses.

2.5.3. Over-Staffing

Compared with other countries, such as USA (20,000), India (50,000), China had more than one million people extension workers in 2001. Between 1996 and 1999, although agricultural extension investment increased rapidly (57%), the increase in the number of staff was even faster (65%). Thus, the proportion of funds used to pay salaries and benefits rose even further.

2.5.4. Poor Quality of Extension Staff

Most extension staff in China has had little or no formal training and education. In 2001 a survey shows that only 10% of extension staff had university level education, and more than 46% had no special training at all. This is in strong contrast to the situation in other countries. Furthermore, even though some staff had benefited from higher education, their special training often did not meet local extension needs or had not been updated to reflect current understanding of agricultural problems or technological opportunities.

2.5.5. Large Amounts of Time that Have to be Spent on Duties Not Related to Extension

Reforms in the 1980s required local extension agencies to (a) allocate staff to other duties unrelated to extension, and (b) to engage in commercial activities in order to generate revenue to maintain or supplement salaries and compensate for the reduced funding for extension. However, because the main commercial activities of the extension workers was (and continues to be) the selling of pesticides and synthetic fertilizers, this led to a conflict of interests. On the one hand they should be encouraging farmers not to overuse fertilizers and pesticides and protect the environment, whilst on the other hand they wish to increase the revenue from the sale of inputs. Moreover, the decentralization of extension staff management led to many extension staff coming under the township governments' administration (whereas previously they came under the county administration). Thus, a large proportion of their salaries are paid by township governments, and this makes it easy for local governments to assign non-extension tasks to the extension people.

2.6. Overuse of Nitrogen Fertilizer Because of the Failure to Take Account of the Agronomic, Economic and Environmental Optimum Application Rate

The agronomic optimum, disregarding the economic return and environmental impacts, is the application rate for maximizing the yield, which varies greatly with the crop variety, yield potential, irrigation/drainage, climate, as well as the fertilizer management technology, etc. Fertilizer trials in China over many years for different crops, soil types, and agro-climates have provided good estimates of how much extra yield can be expected as application rates are increased. For example, according to the analysis of over 2700 field experiments conducted in different regions over the whole China in 1981-1983, the agronomic efficiency of fertilizer N applied to cereals at a rate of 120-150 kg N ha^{-1} was 8.1-11.8 kg yield per ha of N (Zhu and Wen, 1994; Zhu, 1997). However, the response to fertilizer declines with increasing application rates and at very high rates the response is very small and may even be negative. Fertilizer management is an important factor governing the agronomic optimum, and yields under farm conditions. Management levels for small scale fertilizer trials are generally higher than those on most farms, and so the uptake efficiency of fertilizer N by plant may be higher than that in farmers' fields and consequently the agronomic optimum estimated by fertilizer trials will be lower than for normal farm conditions.

The economic optimum N input will be less than the agronomic optimum under both fertilizer trial and farm conditions. This is because diminishing marginal returns at higher yields and fertilizer N inputs will eventually lead to the value of the incremental yield being less than the cost of the extra fertilizer and associated production costs.

Finally, the environmental optimum N input in turn will be comparable or less than the economic optimum because the latter fails to take account of the costs to the public at large of the environmental damage caused by the Npp (Smil, 1996, 1997; Norse et al., 2001). These costs (depending primarily on the extent of overuse) are difficult to estimate and are not known with precision but in the case of rice could be in the range 13-49 billion yuan per year for whole China (Norse et al., 2001).

Consequently, the optimum application rate of fertilizer N for food security and sustainable agriculture is the rate at which the agronomic efficiency, the economic efficiency and the environmental optimum are consistent with each other.

2.7. Over-Fertilization Behaviour of Farmer under Open Market Conditions

In the 1960s China's government recognized the important role of synthetic fertilizers in achieving food security and paid great attention to encouraging the use of fertilizer and ensuring its supply. Fertilizer consumption increased rapidly after the economic reforms of 1978. In 1975 China's farmers applied 70 kilograms per hectare, a level that was about equal to the average fertilizer use intensity of the world. By 2000, however, farmers were applying 280 kilograms per hectare, a level about 3 times the world's average. In terms of fertilizer use intensity, China is ranked fourth in the world after the Netherlands, South Korea and Japan (FAO, 2002).

The rapid growth of fertilizer consumption led to the Chinese government to promote a rapid increase in the production of fertilizer. Pricing policies and direct involvement through state-owned enterprises increased fertilizer production during the 1980s and 1990s. From only

12 million tonnes (measured in nutrient weight) in 1980, China's production of fertilizer increased three fold to 36 million tonnes in 2002, and in 1996 overtook the United States to become the world's leading fertilizer producer.

Although production grew rapidly, consumption rose even faster, and in the 1980s and 1990s, China also became the world's largest importer of synthetic fertilizer. During the 1980s and 1990s China imported an average of 8-9 million tonnes of fertilizer per year, and in the 1990s imports supplied about 25% of annual use. Thus, by the end of the 1990s, China's fertilizer policy and factor endowments made China the world's largest user, producer, and importer of synthetic fertilizer.

Many analysts show that Chinese farmers are overusing fertilizers for most areas, time periods, crops and method of estimation. In the case of farms studied in Jiangsu the ratio of the value of the marginal product of fertilizer to the price of fertilizer is 0 61 to 0.63 (Table 6), that is, the value of extra crop yield produced by additional amounts of fertilizer is only about two-thirds of the cost of the fertilizer so farmers are loosing income by overusing fertilizer.

The results from household level datasets are consistent with those from analyses conducted using the China National Cost of Production Survey Dataset (Table 7). The latter provide strong evidence that from an economic standpoint maize producers are overusing fertilizer by 50 to 75 percent; wheat producers by 33 to 81 percent; and rice producers by 36 to 73 percent. In short, for reasons that need deeper investigation Chinese farmers use fertilizer far in excess of the point of optimal profitability.

The degree of overuse varies regionally in a fairly systematic manner. For example, rice producers in the Yangtze Valley overuse fertilizer by the highest degree. In the 1980s they overused fertilizer by 50 to 65 percent. During the 1990s, the degree of overuse rose marginally to 58 to 70 percent (Table 8). The degree of overuse was less in South China and Southwest China, especially the former, although it increased relatively more in the 1990s than in the Yangste Valley.

Table 6. The rate of total fertilizer overuse by China's farmers in Jiangsu, Hebei and Liaoning provinces, 1995 and 1996

Item	Jiangsu		Hebei and Liaoning	
	FE [a]	FE-IV [b]	FE	FE-IV
Output elasticity of fertilizer	0.10	0.10	0.10	0.08
Ratio of marginal output to fertilizer price	0.63	0.61	0.84	0.69
Fertilizer overuse (%)	44	47	22	35

[a] FE is estimated from models that use household level fixed effects with plot level data;

[b] FE-IV also uses fixed effects, but in addition, the fertilizer variable accounts for both price effects and other unobserved factors .

Table 7. The rate of fertilizer overuse by China's farmers using China National Cost of Production Dataset for different time periods

Corp	Item	1984-1990		1991-2000	
		FE	FE-IV	FE	FE-IV
Corn	Output elasticity of fertilizer	0.11	0.11	0.12	0.08
	Ratio of marginal output to fertilizer price	0.55	0.55	0.52	0.35
	Fertilizer overuse (%)	50	51	56	75
Wheat	Output elasticity of fertilizer[c]	0.17	0.14	0.13	0.06
	Ratio of marginal output to fertilizer price	0.73	0.61	0.51	0.23
	Percentage of fertilizer overuse (%)	33	45	56	81
Rice	Output elasticity of fertilizer[c]	0.13	0.10	0.08	0.05
	Ratio of marginal output to fertilizer price	0.68	0.50	0.44	0.30
	Fertilizer overuse (%)	36	55	61	73

Table 8. The rate of fertilizer overuse by China's rice farmers using National Cost of Production Survey dataset disaggregated by region and time period

Corp	Item	1984-1990		1991-2000	
		FE	FE-IV	FE	FE-IV
Yangste River Valley[a)]	Actual (kg ha^{-1})	792	792	304	304
	Optimal (kg ha^{-1})	398	278	128	90
	Overuse (%)	50	65	58	70
South China[b)]	Actual (kg ha^{-1})	918	918	341	341
	Optimal (kg ha^{-1})	833	683	263	225
	Overuse (%)	9	26	23	34
Southwest China[c)]	Actual (kg ha^{-1})	512	512	233	233
	Optimal (kg ha^{-1})	435	293	135	98
	Overuse (%)	15	43	42	58

[a)]Yangste River Valley includes Anhui, Hubei, Hunan, Jiangsu, Jiangxi, and Zhejiang priovinces;
[b)]South China includes Fujian, Guangdong, and Guangxi provinces;
[c)]Southwest China includes Sichuan, Yunnan, and Guizhou provinces.

The reasons for overuse is that collectives have traditionally pressured farmers more to increase production (for example, to meet food self sufficiency targets) and producers have responded by increasing fertilizer use. It could also be that since farmers in the Yangtze River Valley are more involved in off-farm activities than those in the Southwest, they respond to the rising opportunity cost of labour by applying more fertilizer in a single application in order to reduce labour inputs. Although such rates of application may not be optimal in a strict sense of the definition, and farmers may know that part of the fertilizer will not be used effectively (it may evaporate or be flushed away by the application of surface irrigation), such levels of fertilizer may be rational in that they allow the farmer to focus on his/her higher paid off farm opportunities and neglect farm work. While this may explain the rate of overuse in the Yangste Valley compared with the Southwest, it does not explain the relatively higher rates of overuse when compared to the South. The opportunity cost of farmers in the South must be nearly on par with those in the Yangtze River Valley. There are differences, however, in the institutions in the South and Yangtze River Valley; the higher level of rental transactions in the South may help reduce the levels of inefficiencies since the busiest farmers are more able to rent their land out and do not need to apply excess levels of fertilizer.

The rates of overuse of fertilizer for wheat producers also vary by region. In North China, the heart of China's wheat basket, farmers overuse fertilizer more than those in the rest of China (Table 9). Since the regions in North China have communities that are relatively richer, more industrialized, and more connected to labour markets than most of the regions in the "Rest of China," it could be that North China farmers are overusing fertilizers relatively more than others because their opportunity costs are higher. While these explanations are all plausible, they must remain hypotheses until further studies can be undertaken.

Table 9. The rate of fertilizer overuse by China's wheat farmers using National Cost of Production Survey dataset disaggregated by region and time period

Corp	Item	1984-1990		1991-2000	
		FE	FE-IV	FE	FE-IV
North China [a]	Actual (kg ha^{-1})	962	962	343	343
	Optimal (kg ha^{-1})	533	443	128	60
	Overuse (%)	45	54	63	83
Rest of China [b]	Actual (kg ha^{-1})	581	581	247	247
	Optimal (kg ha^{-1})	435	360	128	60
	Overuse (%)	25	38	48	76

[a] North China includes Hebei, Henan, Shandong, and Shanxi provinces;

[b] rest of China includes Anhui, Gansu, Guizhou, Hubei, Jiangsu, Inner Mongolia, Shaanxi, Sichuan, Xinjiang, and Yunnan provinces.

The overall conclusion of the technical and socio-economic investigations undertaken by the Task Force is that from both bio-physical and economic standpoints Chinese farmers are overusing fertilizer and pesticides by 10—30% for cereals in eastern provinces and up to 50 per cent in the case of intensive vegetables. Such overuse has increased over time and will continue to increase given current trends and policies. The overuse results in low rates of fertilizer use efficiency and high rates of Npp. One of the key factors influencing the overuse is the lack of sound extension advice on fertilizer requirements and methods of application. Better and more frequent extension support by official services or farmer associations, together with improved rural education and training for farmers should lead to more rational and environmentally sustainable fertilizer and pesticide use and greatly reduced Npp.

3. Policy Recommendations to Reduce Non-point Pollution from Agriculture in China

Although the following recommendations are listed separately, they should be formulated and implemented as mutually supporting actions. Moreover, they should be consistent with the overall objectives of : a) income growth and poverty reduction in rural areas; b) integrated rural environment management planning; c) introducing the concept of Environment Impact Assessment into the agricultural planning system; d) applying the concept of the circular economy to agriculture.

3.1. Policy Recommendations

3.1.1. Reassessment of China's Grain Self-Sufficiency Requirements

The growth in grain production has allowed China to become a net rice exporter in recent years. Therefore, at the national level, food security is not a major concern. However, there exist wide differences at the household level. Thus, measures should be taken to tackle micro-level or household level food security. Moreover, past experiences shows that much of the grain security was achieved at the cost of serious environmental damage.

In order to achieve a better balance between food security and environmental quality, we suggest that the target level for grain self-sufficiency should be controlled at 95% thereby reducing the pressures on the environment. According to a recent projection of grain production, imports and exports to 2020, China will import about 50 million tonnes of grain (largely feed grains). Such a level would account for only about 2.6% of world grain production (FAO estimated world production of grain in 2004 to be 1.9 billion tonnes) and less than 22% of world trade (according to FAO the world trade of grain in 2004 was some 230 million tonnes).

The government should undertake several major shifts in its strategy for food security. That is shifting: a.) from a stress on national food security to a stress on household food security in rural and urban areas; b) from grain security to food security; c.) from direct government subsidies to farm households for grain production to a focus on productivity enhancing measures, such as greater investment in agricultural R&D and rural infrastructure.

3.1.2. Change the Regional Structure of Grain Production

It seems possible that Npp could be reduced by changing the regional structure of grain production. The aim would be to lower production in the high yield intensive farming areas where overuse of fertilizer is worst, and raise production elsewhere, particularly in the middle yield areas where soil fertility and productivity of can be improved by irrigation and sound nutrient management. Therefore an agronomic and economic analysis should be undertaken to determine if some of China's future grain production could be moved from the high Npp risk areas of Eastern China (for example, the Taihu lake area) to parts of the central and western provinces where the Npp risk is much less (such as Huang-huai-hai plain). The first action could be to re-assess from an environmental perspective China's development strategies on the current distribution of commercial grain production bases. This analysis must be very thorough because there could be a difficult environmental trade-off between reducing Npp from intensively managed land in Eastern China and increased soil erosion from the more fragile soils of Western and Central China.

3.1.3. Promote Farmers Associations

These play important roles in other countries, but their growth has been restricted in China. Japan introduced the honorary title of Eco-farmer as part of its programme for sustainable agriculture development, and to encourage farmers to protect environment.

International experience with farmers' associations confirms that they can be very effective institutions for bringing small and individual farmers together for marketing, technical training, knowledge transfer and other similar activities. They can also provide credit services to small farmers. Given the small-scale nature of farming in China, and the ineffectiveness of many public extension and farm support activities such institutions are urgently needed. Although government has been trying to promote such organizations only about two percent of rural households currently belong to them. On the other hand, their activities are seriously restricted by their lack of legal identity and the complicated administrative requirements for their establishment. Consequently, a number of policy actions are needed to ensure the healthy development of farmers associations. The actions include the following:

a) Changing the role of government so that it becomes a partnership with farmers. Greater Government support is needed for financial assistance, training, information exchange etc., and should act as one of the catalysts for the establishment of farmers associations;
b) Laws and regulations need to be formulated that recognize the legal rights of farmers associations;
c) Create a favourable environment for the setting up of farmers associations;
d) Allow these organizations to have access to finance or to have the right to form credit unions to help small farmers get access to credit.

The developed area of eastern China may need additional changes in order to allow larger scale farming operations, and the development of specialized farming activities functioning as businesses or corporations.

3.1.4. Raise Environmental Awareness throughout China

First, officials at all levels who are in charge of matters relating to agriculture and environment should study the meaning and scientific method of recent developments in ecologically sound farming. They should be fully aware of relevant national laws and regulations. They need to use all of the available channels of communication, including newspapers, radio and TV to strengthen the distribution of information about agriculturally, ecologically and environmentally sound development in order to widen the participation and ecological consciousness of the whole of society. These actions will require adjustments to the policy framework for sustainable agricultural development regarding taxation, credit availability and market access, etc. and encouragement of investment by Chinese and international enterprises with greater enthusiasm and participation of rural collectives, small farmers and foreign businessman.

3.2. Improvements in Environmental Legislation Recommendations

3.2.1. Tighter Controls on the Discharge of Organic Waste

China's legislation needs to take account of ongoing and future structural changes in the agricultural sector involving shifts from land intensive activities (eg, grain) to labour intensive activities (livestock or horticulture). The rapid growth in intensive livestock production is part of this trend. Improved livestock waste management is becoming a key action for both point and non-point source pollution control.

A number of actions are required to achieve the twin goals of environmental improvement and high-quality and high-yield agriculture. These actions include greater efforts to develop non-polluting agriculture; expand the production of " green food " to promote the application of organic fertilizer; implement the standards for green food production established by the Ministry of Agriculture; control and improve the production process of "organic food"; reduce pesticide pollution from organic fertilizer (such as the veterinary chemicals in livestock's manure).

The government should therefore establish legally enforceable environmental protection requirements for livestock production. These requirements should take account of the livestock carrying capacity of the land; waste storage and disposal needs; buffer zone construction; the potential to increase waste treatment and utilization of livestock's excrements; and the need to limit random discharges of manure.

3.2.2. Promotion of the Recycling of Organic Manure

Rural environmental protection strategies should require the development of comprehensive straw utilization plans that take account of the opportunities for biofuel production to reduce the environmental impact of straw burning and organic fertilizer collection.

The recycling of organic manure needs to be promoted at the regional or local farm scale. This will require policies and regulations about commercial organic fertilizer production and use. These actions should include legislative changes that encourage the development of better techniques for organic fertilizer production; quality standards for commercial organic fertilizers; reduction of nutrient loss during warehousing and transportation; greater efforts to raise the proportion of agricultural wastes used as organic fertilizer. Regulations need to be introduced and implemented to control the application rates and timing of organic fertilizer use (including the ratio between organic manure and inorganic fertilizer) according to climate, soil, and crop condition in order to reduce surface water pollution.

3.2.3. Prevention of Pesticides Pollution

This requires a range of measures covering the production, sale and use of pesticides, and notably the following. National standards are required for the clean production of pesticides using technologies that prevent or reduce emissions and discharges of pollutants and to ensure high and reliable product quality. Farmers need more help on the safe and scientific use of pesticides. Product trademarks need better protection. The system for pesticide residue measurement and quality control of agricultural product needs to be improved and expanded to ensure consumer and environmental safety.

National laws and regulations will have to be strengthened to achieve the above, together with improvements in the powers and functions of environmental protection departments with regard to pesticide security. They will need to be supported by the development and implementation of a monitoring plan for pesticide pollution, and effective controls on the whole process of pesticide production, application, storage and transportation.

The registration and application of pesticides needs to be managed more rigorously, and particularly measures to eliminate highly toxic and stable pesticides, and to develop new pesticides which are environment friendly to (high efficiency, low toxicity pesticides that do not remain in the environment).

The central strategy for pesticide use should be based on the precautionary principle and on integrated control systems. The implementation of this strategy will require the following actions. First, establishment of a plant disease and insect pest forecasting system. Second, popularisation of integrated pest control using biological agents and biological pesticides to reduce the consumption of agricultural chemicals. Third, greater investment in basic research on pest control and applied research on application.

All of the foregoing will be heavily dependent on the reform of the extension service, together with more training and supervision on the safe use of pesticides and greater awareness of the need for safe and well regulated pesticide use.

3.3. Improvement of Technology Delivery Systems

3.3.1. Monitoring the Farmland Quality and Environmental Capacity

The European Commission assesses farmland quality from two aspects: agricultural use and environmental quality. In 1993 the European Commission started national level monitoring every four years. In 1995 the Commission implemented a programme that requires member countries to identify high risk areas for groundwater nitrate, and started to formulate measures to reduce ammonia emission. EU member states, Canada and the USA have for many years had environmental standards for nitrate levels and pesticide residues farmland soil (soil organic content, bio-diversity), food, ground water (nitrates 50 mg L^{-1} and pesticides 0.1 mg L^{-1}).

China should refine or set her own standards for national farmland environmental capacity. These standards should give particular attention to lowering nitrate and pesticide residues, and they should draw on existing, relevant standards and criteria (including those for national soil quality analysis, environmental quality and agricultural product safety (GB, GB/T). These standards could then be used to carry out a nationwide farmland environmental quality survey, and identify and manage high risk areas for non-point pollution.

A national monitoring network for farmland ecology and environmental quality should be established by combining or linking the relevant activities of the Ministry of Agriculture, the Chinese Academy of Sciences and the State Environmental Protection Administration. This monitoring network should be linked to the implementation of rapid residue and quality testing for various green and organic agricultural products as described in recommendation 4.2.1.

3.3.2. Reform the Agricultural Extension System

These recommendations are focused on fertilizer and pesticide use but are relevant to all aspects of agricultural technology delivery. They concern five main actions: a) Reform of the extension system and reduction of the time that extension personnel have to spend on non-extension activities; b) Separation of extension advice from income generation activities, such as selling fertilizer and pesticides; c) Increase the funding for operation and maintenance activities so that extension staff to carry out their proper functions and reach farmers; d) Introduce participatory approaches to give farmers chance to express their own ideas and interests and widen the use of farmer trains farmer techniques; e) Raise the environmental awareness of all extension workers, update their technical knowledge and provide training opportunities that place more emphasis on the environmental consequences of their technical advice to farmers.

3.3.3. Widen the Uptake of Proven High Efficiency Fertilization Technology

Sustainable agricultural development should be based on the principle of "high yield, high input use efficiency, good quality" and "low consumption of natural resources, no pollution". This will require more holistic thinking and increased basic research on agro-ecosystem nutrient cycling and its management; development of new fertilization techniques which are simple and easy to apply, and are suitable under different regional conditions; stronger integration of routine fertilization techniques, and control of non-point nitrogen and phosphorus pollution from agricultural chemicals.

The promotion of sustainable agricultural development could be centred on three lines of action. First, confirmation of the national structure of main crop production (wheat, maize and rice) and the estimation of the correct nitrogen fertilizer rates for different areas using the best scientific methods. Setting up expert decision support systems for precise fertilization at the county scale based on soil nutrient status, transformation processes, crop growth models and the need to decrease fertilizer losses.

Second, promotion of the use of the proven technological measures that lower Npp. Such measures include: optimising the rate of nitrogen fertilizer application using existing recommended technologies; reducing the use of ammonium bicarbonate fertilizers; balanced fertilizer applications tailored to specific soil nutrient (including micronutrients) deficiencies, and cropping systems; deep placement of commercial fertilizers, the use of slow release fertilizers and other forms of precision agriculture; adoption of drip irrigation to raise both water and fertilizer use efficiency; encouraging the use of manures with improved management of the level and timing of manure applications; adoption of no-till and other conservation farming techniques to reduce phosphate and pesticide losses on eroded soil particles; and use of catch or cover crops and buffer strips or diversion drains to capture lost nutrients in natural vegetation or harvestable crops.

Finally development and popularization of new fertilizers. The latter include controlled release ammonium carbonate, controlled release urea, coated urea, foliar sprays and other environment-friendly controlled release fertilizers that reduce the loss of N and P.

3.3.4. Implementation of Comprehensive River Basin Planning and Management

Most European Commission member states have adopted comprehensive river basin management plans that take account of environmental protection policies and regulations.

China needs to adopt similar river basin planning and management approaches if it wishes to control Npp in high risk areas. The focus of such planning should be on raising fertilizer use efficiency, reducing pesticide pollution, and building interception systems to lower the discharge of excess nutrients and pesticides from farmland into surface waters. The plans should also include domestic sewage and livestock waste disposal and recycling projects.

Intercept systems should be established at the river basin level that integrate adjustments in cropping patterns, interplanting, and the maintenance of ground cover to reduce the surface runoff from bare ground and vegetable field as well as more direct measures. The latter can involve ecological interception systems to reduce nutrient loss from farmland, such as the transformation of traditional ditches and modern cement canals into ecological filters, as well setting up ecological interception canals to lower the nutrient content of water draining from farmland. The use of both physical and biological isolation zones and barriers should be considered for high risk pollution plots (particularly vegetable or flower plots) to prevent the spread of N and P.

Measures to improve the quality of water draining from farmland could include relatively conventional agro-ecosystem approaches and new forms of environmental engineering. The latter include low cost ecological techniques to purify water from farmland, for example, artificial wetlands and reed beds. Fishpond waste water could be used for irrigation. Ecological riverbeds could be constructed that use ecological communities and biological processes to repair the vegetation of river corridors and increase the self-purification ability of agro-ecosystems.

REFERENCES

Brown L. 1994. Who will feed China? *World Watch*, 7(5): 66-76.

Conway G. 1994. Sustainable agriculture for a food secure world, CGIAR, Washington, D. C. 1-64.

Dong K Y. 1998. Reclamation and Environment pollution of Wastes from Livestock and Poultry. *Agro-environmental Protection*, 17(6):281-283. (in Chinese).

Duan S W, Zhang S, Huang H Y. 2000. Transport of dissolved inorganic nitrogen from the major rivers to estuaries in China. *Nutrient Cycling in Agroecosystems*, 57(1): 13-22.

Editor Committee of China Agricultural Yearbook. 1991-2002. China Agricultural Yearbook. Chinese Agriculture Press, Beijing, China. (in Chinese).

Gao D, Chen T B, Liu B, et al..2006. Releases of pollutants from poultry manure in China and recommended strategies for the pollution prevention. *Geographical Research*, 35(2): 311-319.

International Fdertilizer Industry Asscociation (IFA), Food and Agriculture Organization of the United Nationations (FAO). 2001. Globa 1 estimates of gaseous emissions of NH_3, NO and N_2O from agricultural land. ISBN 92-5-104689-1, Rome. 1-106.

Jiang G H, Wang W K, Yang X T et al. 2002. Analysis of nitrate pollution of groundwater of Guanzhong Basin and countermeasures. *Water Resources Protection*, (2): 6-8. (in Chinese).

Li G B, Yin C Q, Zhou H D. 2001. Three-lake water problem of China and its countermeasures and management. *Water Problem Forum*, (3): 36-39. (in Chinese).

Lu D Q, Dong Y A, Sun B H. 1998. Study on effect of nitrogen fertilizer use on environment pollution. *Plant Nutrition and Fertilizer Sciences*, 4(1): 8-15. (in Chinese).

Ma W Q, Mao D R, Zhang F S. 2000. The problems in fertilization and measurements of preventing them in protective vegetable ground in Shandong. In Li, X.L., Zhang, F.S., Mi G. H.(eds.).Fertilizing for sustainable production of high quality vegetables. Chinese Agricultural University Press, Beijing. 41-47. (in Chinese)

Norse D, Li J, Jin L et al. 2001. Environmental Costs of Rice Production in China: Lessons from Hunan and Hubei, Aileen International Press, Bethesda, MD, USA. 93.

Shen R P, Sun B, Zhao Q G. 2005. Spatial and Temporal Variability of N, P and K Balances in Agroecosystems in China. *Pedosphere*, 15(3): 347-355.

Smil V. 1997. China's environment and security: simple myths and complex realities. SAIS Review, 17:107-126.

Smil V. 1996. Environmental Problems in China: Estimates of Economic Costs. East-West Center, Honolulu, HI, 62 pp.

State Environment Protection Agency (SEPA). 1999. 1998 Report on the State of Environment in China. *Enviroenment Protection*, (7):3-9. (in Chinese).

State Environment Protection Agency (SEPA). 2004. 2003 Report on the State Environment in China. *Enviroenment Protection*, (7):3-17 (in Chinese).

Sun B, Chen D, Li Y, Wang X. 2008. Nitrogen leaching in an upland cropping system on an acid soil in subtropical China: lysimeter measurements and simulation. *Nutrient Cycling in Agroecosystems*, 81(3): 291–303.

Xin G X, Cao Y C, Shi S L et al. 2001. Pollution Source of N and denitrification in the water body in Tai Lake region. Science in China (B), 31(2): 130-137. (in Chinese)

Xin G X, Zhu Z L. 2000. An assessment of N loss from agricultural fields to the environment in China. Nutr Cycl Agroecosyst. 57(1): 67-73.

Yang L Z, Sun B. 2008. Cycling, Balance and Management of Nutrients in Agroecosystems in China. Science Press of China, Beijing, 310pp.

Yuan X Y. 2000. Primary appraisal of pollution for lakes of China. *Volcanology and Mineral of Resources*, 21(2): 128-136. (in Chinese)

Zhang F S. 1999. Some consideration to the improvement of nutrient resources utilization efficiency. In: Soil Science Society of China (ed.), Soil Science Towards 21th Century. Proceedings of 9th National Congress of Soil Science Society of China, Nanjing, 42-48. (in Chinese).

Zhang W L, Tian Z X, Zhang N et al. 1995. Investigation of nitrate pollution in ground water due to nitrogen fertilization in agriculture in North China. *Plant Nutrition and Fertilizer Sciences*, 1(2): 80-87.

Zhang Y, Liu X J, Zhang F S, et al. 2006. Spatial and temporal variation of atmospheric nitrogen deposition in North China Plain. *Acta Ecologica Sinica*, 26(06): 1633-1639.

Zhu Z L. 1997. Fate and management of fertilizer nitrogen in agroecosystems. In Zhu Z L, Wen Q X and Freney J R (eds), Nitrogen in Soils of China. Kluwer Academic Publishers: Dordrecht. 239-279.

Zhu Z L. 2003. Fertilizer management strategies for the harmonization of agriculture development with environment protection. *Publication of Chinese Academy of Sciences*, (2): 89-93. (in Chinese)

Zhu Z L, Norse D, Sun B. 2006. Policy for Reducing Non-point Pollution from Crop Production in China. China Environmental Science Press, Beijjing, 287pp

Zhu Z L, Sun B. Study on the countermeasures to control non-point pollution of agriculture in China. *Environmental Protection*, 2008, (8): 4-6.

Zhu Z L, Wen Q X. 1994. Soil Nitrogen of China. Jiangshu Science and Technology Press, Nanjing, China. 1-303. (in Chinese).

In: Pollution in China
Editor: Michael I. Chang

ISBN: 978-1-61122-022-3
©2011 Nova Science Publishers, Inc.

Chapter 8

HEAVY METAL CONTAMINATION OF AGRONOMIC CROPS GROWN ON THREE RECLAIMED MINE WASTELANDS IN SOUTH CHINA AND IMPLICATIONS FOR ECOLOGICAL RESTORATION*

***Ming-Shun Li*[1*∞], *Yan-Ping Lai*[2] and *Shichu Liang*[2]**

[1] School of Environment and Resources, Guangxi Normal University, Guilin 541004, China

[2] School of Life Sciences, Guangxi Normal University, Guilin 541004, China

ABSTRACT

Agronomic crops grown on the reclaimed metal-mined wastelands are a pathway for toxic pollutants entering the human food chain. Agricultural rehabilitation of mine spoils in China is pretty common and its effect has been largely overlooked. Extensive sampling of the edible crops and associated soils have been conducted for the three typical manganese mine wastelands (Bayi, Lipu and Pingle) in Guangxi, south China and heavy metal contamination of crops was assessed against China Food Safety Standards. Simple pollution index (Pi) assessment indicated no Zn (except tea) and Cu pollution among these crops, but heavy pollution of Pb, Cd and Cr was found. Composite pollution index (Nemerow index, P_N) showed 36 crops from 41 were heavily polluted with heavy metals. Peanut, soybean, Chinese chestnut, persimmon, cassava, mandarin and sugarcane were the most severely contaminated crops. Consumption of these crops may pose a health risk for humans. Crops tended to have a higher Cd accumulation (as indicated by Biological Accumulation Factor) in edible parts, thus Cd is the most important food safety threat. In terms of China Soil Quality Standard (class II), the minesoils contained much higher Cd and Cr levels, not suitable for agricultural plantation. Simple reclamation for crop plantation on minesoils is legally untenable and must be strictly controlled by the local

* A version of this chapter was also published in Coal Mining: Research, Technology and Safety, edited by Gerald B. Fosdyke. It was submitted for appropriate modifications in an effort to encourage wider dissemination of research.

∞ Corresponding author: Tel: 86 773 5829507, Fax: 86 773 5846201, msli@mailbox.gxnu.edu.cn governments. In addition, more diverse restoration goals with lower environmental risk should be encouraged for the mine wastelands in South China.

Keywords: heavy metal contamination, agronomic crops, Mn mine wasteland, simple pollution index, composite pollution index, ecological restoration, Guangxi.

INTRODUCTION

Food safety is fundamental to human survival and health. Recently, adverse publicity about China's contaminated food exports (e.g., rejected shipments of vegetables, tea, dumplings, shrimp, and poultry) reflected a gap between international and China's food safety standards. Also, a series of food safety incidents in the domestic market (e.g., respiratory diseases, acute pesticide residual poisons, and sudan yellow dye) prompted the Chinese government to raise domestic food safety standards and implement inspection and testing systems for consumer products and agricultural commodities. Metallic elements, either essential or nonessential, are potentially harmful food safety threats because they are generally invisible, accumulative (slow), and irreversible in effect. Of these elements, the most important in terms of food-chain contamination are arsenic (As), cadmium (Cd), mercury (Hg), lead (Pb), and selenium (Se) (McLaughlin et al., 1999). Except for Se, these toxic metals enter soils through anthropogenic sources, such as wastewater irrigation, sludge application, chemical fertilizer usage, ore extraction and smelting, industrial wastes, and atmospheric deposition.

It is reported widely that about 1/5 of the cultivated land (20 Mha) in China were contaminated with heavy metals (Sun et al., 2005), among which Cd and Pb were the main contaminants. Heavy metal contamination of foods has attracted great concern in China, especially in the big cities. Assessment of heavy metal pollution of paddy fields, kale yard soils, or orchard soils had been conducted extensively in the outskirts of Beijing, Shanghai, Guangzhou, Tianjing, Hanzhou, Fuzhou, Shenyang, Zhenzhou, Xi'an, Chendu, Chongqing, and Nanning (Liu & Chen, 2004), many of which the pollution of crops also was evaluated. For instance, in the notorious Zhangshi Irrigation Area of north China, the mean Cd level in its unpolished rice reached 1.06 mg/kg while the maximum allowed level was only 0.2 mg/kg (Chan & Shi, 2001). In the Pearl River Delta region, the world's manufacturing base which provides over 80% of the vegetables for Hong Kong, 40% of the farmland and vegetable fields were contaminated with multiple metals, and nearly 1/3 of the main-consumed vegetables were contaminated by metals especially Cd and Pb (Huayi net news, March 6, 2007).

Although the heavy metal contamination of staple food and their associated agricultural lands was really alarming, pollution of the minelands largely has been overlooked in China because they are usually remote to the densely populated urban areas. Mining is the second principal source of heavy metal contamination in soil after sewage sludge (Singh et al., 2005). As a primary contributor to the rapid economic boost of China, mining industry generated vast areas of wastelands (about 3.2 Mha) and caused serious environmental pollution (Li, 2006). A conservative estimate of economic loss (direct and indirect) resulted from mining pollution is nearly 40 billion RMB *yuan* in China each year (Liu & Shu, 2003). In addition, problems existed for the restoration practices in China because many of these wastelands were reclaimed for planting agronomic crops because China's cultivated lands are in serious shortage. In fact, the input of toxic metals into food web generates another potentially more harmful problem than the wasteland itself.

China has the world's second largest Mn-ore reserve. Guangxi *Zhuang* Autonomous Region, adjoining Guangdong Province in the east and bordering Vietnam to the southwest, is one of the most well-developed karst areas on earth. Meanwhile, mining is a pillar industry in Guangxi with Mn and Sn mining ranking first in China. Currently, there are over 6800 mines in operation (96% of the mines were private or collective-owned), destroying a total area of about 600,000 ha (Li, 2005). Due to the extreme shortage of surface soil and economic backwardness, agricultural reclamation of mine wastelands is more common in Guangxi than in other parts of China. The major edible plants grown include Chinese chestnut, sugarcane, peanut, mandarin, orange, plum, peach, persimmon, Taiwan green Jujube, shaddock, medical herbs, tea trees, and other vegetables such as Chinese cabbage and Chinese radish. Worries arose because there were usually no protective treatments in place before planting and no monitoring of toxic metals was conducted or source of food was indicated before they entered the market. This study, based on the investigation on the three typical Mn mine wastelands in Guangxi, aims to assess the heavy metal contamination of grown crops against China Food Safety Standards, and furthermore, the restoration practice is discussed in hope to provide insight to the proper rehabilitation of metal-mined wastelands in South China.

Materials and Methods

The Study Site

The three Mn mines (Bayi, Lipu and Pingle) in Guangxi all employed opencast mining which ore extraction began in the late 50s. Lipu and Pingle mines were both medium-sized, about 105-115 km southeast from Guilin, the world's famous karst scenic spot (Fig. 1). Bayi mine was once one of the three largest Mn mines in China, situated in the central Guangxi. Due to the ore depletion, large excavations have ceased for the three mines, but some private miners were digging for residuals at the margin. These areas belong to the middle subtropical monsoon climatic zone. The regional vegetation is the typical subtropical evergreen broadleaf forest. Landform of these mines is basically hilly land and zonal soil is loess.

Large reclamation efforts (reclamation area > 10 ha) were made for these mine wastelands: Huge lands were planted with sugarcane and tea trees in Bayi while Chinese chestnuts were planted in Lipu and peach trees in Pingle (Yang et al., 2007). Other agronomic plants (e.g. peanuts) were grown in smaller scale or even small patches (e.g. vegetables) for family consumption. In addition to the edible crops, other reclamations for nursery, pulpwood, charcoal and medical herbs existed while some wastelands just lay fallow. In terms of the overall vegetation coverage, Lipu (90%) is the highest, and Pingle (30%) the lowest with Bayi (75%) in-between.

Figure 1. Map of Guangxi, showing locations of the three Mn mine wastelands: Bayi, Lipu, and Pingle.

Sample Collection and Analysis

From November 2004 to October 2006, extensive ecological surveys and sampling have been carried out in the three mine wastelands respectively. The major edible crops were sampled at their appropriate seasons, each sample comprising 5-6 multipoint subsamples. Meanwhile, the associated top soils (0-20 cm) were collected for metal determination. Both plant and soil samples were taken in triplicates and sealed with polythene bags and transported into laboratory.

Plant samples were gently washed with tap water, and rinsed three times with deionized water. Samples were air-dried and weighed, first dried at 105℃ for 30 min, and then at 70℃ to constant weight. Dried plant materials were ground into fine powder. Soil samples were air-dried, homogenized and sieved through a 2-mm screen, then pulverized and passed through a 0.154-mm nylon sieve. Soil samples were digested with concentrated HCl + concentrated HNO_3 + HF + $HClO_4$ (10:5:5:3, v/v), and plant tissues digested with concentrated HNO_3 + $HClO_4$ (20:3-5, v/v). The total metal concentrations (Cd, Cr, Cu, Pb, Zn and Mn) in digest were determined with flame atomic absorption spectrophotometer. Quality assurance of metal determination was executed using rate of recovery of the added standard amount of metal into the digested solution, and the recovery rates for these measurements were within 89 - 106%. Statistical analyses were performed using SPSS 12 for Windows.

Pollution Assessment

Assessment Method

The simple pollution index (Pi) and composite pollution index (Nemerow index, P_N) were employed to assess the pollution degree of the edible crops. Pi considers single metal pollution separately whereas P_N, integrating the mean Pi with the extreme pollution scenario, is a comprehensive indication of pollution (Chen, 2006):

Pi = Ci / Si

where Ci represents the concentration of heavy metal *i* in plant tissue while Si indicates the relevant standard value for this metal (see the next); and

$$P_N = \sqrt{\frac{P_{i(ave)}^2 + P_{i(max)}^2}{2}}$$

where $P_{i(ave)}$ is the average of Pi of metals, and $P_{i(max)}$ denotes the maximum value among Pi.

Assessment Criteria and Pollution Grading

The maximum allowable levels of contaminants in foods of China were used as the assessment criteria. Table 1 listed the maximum levels of Zn, Pb Cr, Cu and Cd in foods relevant to this study. There are no values stipulated for Mn, thus Pi and P_N for Mn were not calculated. Based on Pi and P_N values, heavy metal contamination were classified into different grades and the corresponding pollution levels were given in Table 2.

Table 1. Maximum allowable level of heavy metals in foods as assessment criteria

Food category	Allowable metal level ($mg \cdot kg^{-1}$ FW)				
	Zn	Pb	Cr	Cu	Cd
Beans	100	0.2	1.0	20	0.2 (soybean) 0.5 (peanut)
Potatoes & tubes	50	0.2	0.5	10	0.1
Fruits	5	0.1	0.5	10	0.05
Tea	100[a]	5	5[b]	60[c]	1[b]
Vegetables	20	0.3 (leaf) 0.1 (non-leaf)	0.5	10	0.1 (tuber) 0.2 (leaf) 0.05 (others)
Standards[d]	GB13106-1991	GB2762-2005	GB2762-2005	GB15199-1994	GB2762-2005

[a] Food categories were not exhaustive, and only relevant values related to this study were listed. For tea, only maximum Pb level was given. For Zn, the maximum level of foods was used in the assessment.

[b] For Cr and Cd levels in tea, the values are from the standard NY659-2003 issued by the Ministry of Agriculture, China.

[c] For Cu level in tea, the value is from the standard NY5017-2001 issued by the Ministry of Agriculture, China.

[d] These standards were issued by the Ministry of Health, China.

Table 2. Pollution index grading and corresponding pollution level of heavy metals

Grade	Simple pollution index (Pi)	Pollution level	Composite pollution index (P_N)	Pollution level
1	$P_i < 1$	Unpolluted	$P_N \leq 0.7$	Unpolluted
2	$1 \leq P_i < 2$	Slight	$0.7 < P_N \leq 1$	Warning
3	$2 \leq P_i < 3$	Medium	$1 < P_N \leq 2$	Slight
4	$3 \leq P_i$	Heavy	$2 < P_N \leq 3$	Medium
5	-	-	$3 < P_N$	Heavy

Bioaccumulation Factor

The Bioaccumulation Factor (BAF) of a plant is the heavy metal concentration in plant tissue (dry weight) divided by the same metal concentration in soil (Li et al., 2007), and this expression is often used to evaluate the plant's uptake capacity of the metal from soil. Low accumulating crops (cultivars) of toxic metals are better choices for phytoremediation of metal-contaminated site.

RESULTS

Heavy Metals in Agronomic Crops

There were 9 crops from Bayi, 19 from Lipu and 13 from Pingle mine wastelands included in this study, and only the edible parts for human being were analyzed for heavy metals. The heavy metal contents in the edible parts of crops are presented in Table 3, and the summarized data in Table 4. If we consider each crop as an individual sample, the heavy metal levels in crops were generally in the order of Mn > Zn > Pb > Cr ~ Cu > Cd. There were great variations of metal content among different crops for the same metal and among different wastelands (Table 3). The highest values for Zn, Pb, Cr, Cu, Cd and Mn were in tea (Bayi), peanut (Pingle), tea (Bayi), tea (Bayi), peanut (Lipu) and tea (Bayi), respectively for each metal, and the lowest were in orange (Bayi), watermelon (Bayi), watermelon (Bayi), orange (Lipu), Nachi pear (Lipu) and Nachi pear (Lipu).

Overall, median values were all lower than the mean especially for Mn (Table 4) because median value eliminates the effect of extreme figures in a data set. Compared with the heavy metal levels of crops grown on the normal farmlands of China (Zhong et al., 2001), Zn and Cu levels of the crops on these minelands were well within the range, but Pb, Cd or Cr level, regardless of mean or median, was higher than the upper limit, indicating a potential multi-metal pollution.

Table 3. Heavy metal contents (mean ± SE, *n*=3) of agronomic crops on the three Mn mine wastelands in Guangxi

Crop	Scientific name	Edible part	Heavy metal content ($mg \cdot kg^{-1}$ FW)					
			Zn	Pb	Cr	Cu	Cd	Mn
Bayi mineland: 9 edible crops								
Peanut	*Arachis hypogaea*	earthnut	21.44±0.86	4.80±0.47	1.45±0.38	6.37±0.32	0.52±0.11	17.14±0.97
Soybean	*Glycine max*	seed	37.53±1.02	8.01±0.46	3.76±0.81	6.10±0.38	1.03±0.18	17.74±0.74
Sweet potato	*Ipomoea batatas*	tuber	3.12±0.69	3.24±0.09	2.41±0.55	1.81±0.13	0.39±0.06	11.41±3.25
Sweet potato	*Ipomoea batatas*	leaf	4.23±0.31	1.95±0.06	1.48±0.05	1.97±0.08	0.33±0.10	25.52±2.66
Cassava	*Manihot esculenta*	tuber	6.01±0.44	2.31±0.35	1.94±0.39	1.96±0.18	0.74±0.15	3.5±0.06
Orange	*Citrus sinensis*	fruit	1.03±0.08	0.66±0.09	0.27±0.12	0.39±0.03	0.18±0.02	0.78±0.24
Watermelon	*Citrullus lanatus*	fruit	1.24±0.13	0.32±0.02	0.11±0.01	0.16±0.01	0.06±0.01	1.55±0.03
Sugarcane	*Saccharum sinensis*	shoot	3.19±0.81	2.09±0.29	1.92±0.31	0.74±0.04	0.54±0.06	24.4±1.15
Tea	*Camellia sinensis*	leaf	96.02 ±21.39	13.01±0.54	9.28±1.69	9.71±0.16	0.76 ±0.13	1768.45±136.56
Lipu mineland: 19 edible crops								
Peanut	*Arachis hypogaea*	earthnut	20.46±1.52	5.73±1.36	3.09±0.35	3.84±0.03	2.76±0.06	23.62±1.08
Soybean	*Glycine max*	seed	20.46±0.51	8.96±2.00	2.34±0.19	6.24±0.18	2.30±0.09	30.14±1.71
Sweet potato	*Ipomoea batatas*	tuber	5.07±0.24	6.26±1.98	2.46±0.39	0.04±0.06	1.51±0.62	2.65±0.86
Sweet potato	*Ipomoea batatas*	leaf	2.93±0.07	1.04±0.12	1.07±0.18	0.76±0.01	0.08±0.00	8.83±0.06
Cassava	*Manihot esculenta*	tuber	7.41±0.47	7.33±1.75	2.32±0.22	2.39±0.07	1.79±0.11	1.49±0.09
Orange	*Citrus sinensis*	fruit	1.94±0.32	5.39±0.86	1.01±0.11	0.01±0.09	0.24±0.03	1.59±0.64
Sugarcane	*Saccharum sinensis*	shoot	3.65±0.21	7.34±1.87	2.09±0.09	2.22±0.06	1.80±0.09	12.18±0.09
Nachi pear	*Pyrus pyrifolia*	fruit	2.12±0.92	0.66±0.18	2.02±0.29	0.85±0.02	0.01±0.01	0.66±0.09
Persimmon	*Diospyros kaki*	fruit	2.99±0.18	5.88±1.25	2.47±0.53	0.18±0.04	0.72±0.04	7.58±0.51
Peach	*Amygdalus persica*	fruit	2.86±0.32	1.50±0.08	1.06±0.10	0.74±0.05	0.06±0.01	1.37±0.08
Chinese Chestnut	*Castanea mollissima*	nut	9.58±1.64	11.05±0.81	2.94±0.79	3.45±0.31	1.89±0.17	66.27±6.97
Shallot	*Allium fistulosum*	shoot + leaf	4.26±0.43	1.70±0.47	1.34±0.37	0.56±0.00	0.36±0.01	21.18±5.11

Table 3. (Continued)

Crop	Scientific name	Edible part	Heavy metal content ($mg \cdot kg^{-1}$ FW)					
			Zn	Pb	Cr	Cu	Cd	Mn
Garlic	*Allium sativum*	shoot + leaf	4.50±0.78	2.29±0.71	1.19±0.04	0.78±0.04	0.64±0.04	19.72±2.10
Capsicum	*Capsicum annuum*	fruit	8.82±1.01	8.21±1.35	4.87±1.42	2.78±0.31	0.92±0.05	8.21±2.07
Eggplant	*Solanum melongena*	fruit	1.18±0.05	0.65±0.05	1.07±0.20	0.43±0.01	0.03±0.00	13.09±4.22
Chinese radish	*Raphanus sativus*	tuber	2.45±0.21	0.88±0.23	0.63±0.20	0.26±0.01	0.13±0.01	2.85±0.18
Cowpea	*Vinga unguiculata*	fruit	5.13±0.04	2.49±0.65	0.93±0.14	0.52±0.16	0.43±0.06	50.15±0.96
Endive	*Cichorium endivia*	leaf	5.20±0.24	0.83±0.11	0.75±0.09	0.42±0.02	0.12±0.00	6.37±0.24
Lettuce	*Lactuca sativa*	leaf	3.30±0.18	0.52±0.15	0.33±0.02	0.34±0.05	0.20±0.00	5.86±0.38
Pingle mineland: 13 edible crops								
Peanut	*Arachis hypogaea*	earthnut	15.39±0.26	16.16±2.83	6.62±0.73	2.74±0.11	0.74±0.04	9.64±0.09
Chinese Chestnut	*Castanea mollissima*	nut	9.17±1.50	3.50±0.35	3.43±0.32	5.30±0.25	0.98±0.05	69.82±2.02
Soybean	*Glycine max*	seed	18.13±0.08	4.23±0.62	5.47±0.58	5.21±0.29	0.04±0.02	29.62±0.67
Cowpea	*Vinga unguiculata*	fruit	6.85±0.19	1.02±0.06	2.32±0.17	0.70±0.01	0.06±0.00	60.1±1.73
Sweet potato	*Ipomoea batatas*	tuber	4.78±0.17	9.91±0.26	3.69±0.11	1.20±0.13	0.78±0.03	40.47±6.69
Cassava	*Manihot esculenta*	tuber	6.06±0.78	10.46±3.67	1.18±0.10	0.54±0.10	0.68±0.04	17.51±3.24
Mandarin	*Citrus reticulata*	fruit	1.49±0.05	5.90±0.40	2.23±0.04	0.34±0.05	0.48±0.05	5.02±0.33
Persimmon	*Diospyros kaki*	fruit	1.80±0.43	8.62±0.07	0.97±0.17	0.28±0.07	0.55±0.09	21.22±6.41
Peach	*Amygdalus persica*	fruit	1.56±0.11	0.66±0.09	0.96±0.14	0.32±0.01	0.01±0.00	1.29±0.03
Taro	*Colocasia esculenta*	tuber	19.28±2.99	6.76±1.61	3.78±0.00	1.16±0.22	0.73±0.04	160.79±4.10
Ginger	*Zingiber officinale*	tuber	4.69±0.12	2.93±0.27	1.51±0.11	0.58±0.04	0.32±0.02	334.09±41.68
Tomato	*Lycopersicon esculentum*	fruit	1.27±0.04	0.54±0.04	1.19±0.08	0.45±0.14	0.03±0.00	1.77±0.03
Capsicum	*Capsicum annuum*	fruit	3.25±0.16	3.11±0.27	1.60±0.23	1.28±0.08	0.38±0.05	16.83±2.43

Table 4. Summarized analyses of heavy metal concentrations in agronomic crops on the three mine wastelands in Guangxi

	Heavy metal concentration ($mg \cdot kg^{-1}$)					
	Zn	Pb	Cr	Cu	Cd	Mn
Bayi mineland						
Maximum	96.02	13.01	9.28	9.71	1.03	1768.45
Minimum	1.03	0.32	0.11	0.16	0.06	0.78
Mean	19.31	4.04	2.51	3.25	0.51	232.39
Median	4.23	2.31	1.92	1.96	0.52	17.44
Number of crops	9	9	9	9	9	9
Lipu mineland						
Maximum	20.46	11.05	4.87	6.24	2.76	66.27
Minimum	1.18	0.52	0.33	0.01	0.01	0.66
Mean	6.02	4.14	1.79	1.41	0.84	14.94
Median	4.26	2.49	1.34	0.74	0.43	8.21
Number of crops	19	19	19	19	19	19
Pingle mineland						
Maximum	19.28	16.16	6.62	5.3	0.98	334.09
Minimum	1.27	0.54	0.96	0.28	0.01	1.29
Mean	7.21	5.68	2.69	1.55	0.44	59.09
Median	4.78	4.23	2.23	0.7	0.48	21.22
Number of crops	13	13	13	13	13	13
Overall						
Maximum	96.02	16.16	9.28	9.71	2.76	1768.45
Minimum	1.03	0.32	0.11	0.01	0.01	0.66
Mean	9.31	4.61	2.23	1.86	0.64	72.78
Median	4.5	3.24	1.92	0.76	0.48	14.96
Number of crops	41	41	41	41	41	41
Range in crops of China ($mg \cdot kg^{-1}$)	2.547~26.33	0.01~3.265	0.069~0.651	0.384~5.86	0.012~0.319	-

Pollution Assessment of Agronomic Crops

The pollution indices (Pi and P_N) of the crops and the corresponding pollution levels are shown in Table 5. For the five toxic metals studied (Mn was not assessed since no criteria of food safety is available), basically no Zn (except tea) or Cu pollution existed in these crops. Almost all crops were polluted with Pb, Cd and Cr. The pollution patterns of heavy metals were consistent throughout the three wastelands.

From a comprehensive consideration of heavy metals, 36 crops (87.8%) of 41 were heavily polluted; only 3 (orange in Bayi, sweet potato leaf in Lipu and endive in Lipu) were polluted moderately, and 2 (watermelon in Bayi and lettuce in Lipu) were slightly polluted (Table 5). If $P_N > 9$ (three times the 'heavy' limit) was regarded as extremely heavy pollution, the percentage of this class represented 63.4% of the total edible crops. Consumption of these food crops may pose a great health risk for humans.

Table 5. Pollution assessment of the edible crops on the three mine wastelands in Guangxi

Crop	Pollution index of metals in crops by region					
	Pi (Zn)	Pi (Pb)	Pi (Cr)	Pi (Cu)	Pi (Cd)	P_N
Bayi						
Peanut	0.2 unpolluted[a]	24 heavy	1.5 slight	0.3 unpolluted	1 slight	17.4 heavy
Soybean	0.4 unpolluted	40.5 heavy	3.8 heavy	0.3 unpolluted	5.2 heavy	29.5 heavy
Sweet potato (tuber)	0.1 unpolluted	11.6 heavy	3.9 heavy	0.2 unpolluted	7.4 heavy	8.8 heavy
Sweet potato (leaf)	0.2 unpolluted	6.5 heavy	3 heavy	0.2 unpolluted	1.7 slight	4.9 heavy
Cassava	0.1 unpolluted	16.2 heavy	4.8 heavy	0.2 unpolluted	3.9 heavy	12.0 heavy
Orange	0.2 unpolluted	3.3 heavy	0.5 unpolluted	0 unpolluted	3.6 heavy	2.8 medium
Watermelon	0.2 unpolluted	1.6 slight	0.2 unpolluted	0 unpolluted	1.2 slight	1.2 slight
Sugarcane	0.6 unpolluted	20.9 heavy	3.8 heavy	0.1 unpolluted	10.8 heavy	15.6 heavy
Tea	4.8 heavy	2.6 medium	18.6 heavy	0.97 unpolluted	3.8 heavy	13.9 heavy
Lipu						
Peanut	0.2 unpolluted	28.7 heavy	3.1 heavy	0.2 unpolluted	5.5 heavy	21.0 heavy
Soybean	0.2 unpolluted	44.8 heavy	2.3 medium	0.3 unpolluted	11.5 heavy	32.8 heavy
Sweet potato (tuber)	0.1 unpolluted	31.3 heavy	4.9 heavy	0 unpolluted	15.1 heavy	23.3 heavy
Sweet potato (leaf)	0.1 unpolluted	3.5 heavy	2.1 medium	0.1 unpolluted	0.4 unpolluted	2.6 medium
Cassava	0.1 unpolluted	36.6 heavy	4.6 heavy	0.2 unpolluted	17.9 heavy	27.2 heavy
Orange	0.4 unpolluted	53.9 heavy	2 medium	0 unpolluted	4.8 heavy	39.1 heavy
Sugarcane	0.7 unpolluted	73.4 heavy	4.2 heavy	0.2 unpolluted	36 heavy	54.4 heavy
Nachi pear	0.4 unpolluted	6.6 heavy	4 heavy	0.1 unpolluted	0.2 unpolluted	4.9 heavy
Persimmon	0.6 unpolluted	58.8 heavy	4.9 heavy	0 unpolluted	14.4 heavy	43.0 heavy
Peach	0.6 unpolluted	15 heavy	2.1 medium	0.1 unpolluted	1.2 slight	10.9 heavy
Chinese Chestnut	1.9 slight	110.5 heavy	5.9 heavy	0.3 unpolluted	37.8 heavy	81.2 heavy
Shallot	0.2 unpolluted	5.7 heavy	2.7 medium	0.1 unpolluted	1.8 slight	4.3 heavy
Garlic	0.2 unpolluted	7.6 heavy	2.4 medium	0.1 unpolluted	3.2 heavy	5.7 heavy
Capsicum	0.4 unpolluted	82.1 heavy	9.7 heavy	0.3 unpolluted	18.4 heavy	60.1 heavy
Eggplant	0.1 unpolluted	6.5 heavy	2.1 medium	0 unpolluted	0.5 unpolluted	4.8 heavy
Chinese radish	0.1 unpolluted	8.8 heavy	1.3 slight	0 unpolluted	2.5 medium	6.5 heavy
Cowpea	0.3 unpolluted	24.9 heavy	1.9 slight	0.1 unpolluted	8.6 heavy	18.3 heavy
Endive	0.3 unpolluted	2.8 medium	1.5 slight	0 unpolluted	0.6 unpolluted	2.1 medium
Lettuce	0.2 unpolluted	1.7 slight	0.7 unpolluted	0 unpolluted	1 slight	1.3 slight

Table 5. Continued

Crop	Pollution index of metals in crops by region					
	Pi (Zn)	Pi (Pb)	Pi (Cr)	Pi (Cu)	Pi (Cd)	P_N
Pingle						
Peanut	0.2 unpolluted	80.8 heavy	6.6 heavy	0.1 unpolluted	1.5 slight	58.5 heavy
Chinese Chestnut	1.8 slight	35 heavy	6.9 heavy	0.5 unpolluted	19.6 heavy	26.3 heavy
Soybean	0.2 unpolluted	21.2 heavy	5.5 heavy	0.3 unpolluted	0.2 unpolluted	15.5 heavy
Cowpea	0.3 unpolluted	10.2 heavy	4.6 heavy	0.1 unpolluted	1.2 slight	7.6 heavy
Sweet potato (tuber)	0.1 unpolluted	49.5 heavy	7.4 heavy	0.1 unpolluted	7.8 heavy	36.2 heavy
Cassava	0.1 unpolluted	52.3 heavy	2.4 medium	0.1 unpolluted	6.8 heavy	38.0 heavy
Mandarin	0.3 unpolluted	59 heavy	4.5 heavy	0 unpolluted	9.6 heavy	43.0 heavy
Persimmon	0.4 unpolluted	86.2 heavy	1.9 slight	0 unpolluted	11 heavy	62.6 heavy
Peach	0.3 unpolluted	6.6 heavy	1.9 slight	0 unpolluted	0.3 unpolluted	4.8 heavy
Taro	0.4 unpolluted	33.8 heavy	7.6 heavy	0.1 unpolluted	7.3 heavy	24.9 heavy
Ginger	0.2 unpolluted	29.3 heavy	3 heavy	0.1 unpolluted	3.2 heavy	21.3 heavy
Tomato	0.1 unpolluted	5.4 heavy	2.4 medium	0 unpolluted	0.6 unpolluted	4.0 heavy
Capsicum	0.2 unpolluted	31.1 heavy	3.2 heavy	0.1 unpolluted	7.5 heavy	22.8 heavy

[a] Right to the pollution index is pollution level according to the Table 2.

Heavy Metals in Soils and Crop Accumulation

A total of 21 crop-associated top soil samples (0-20 cm) from Bayi, 34 samples from Lipu, and 34 from Pingle were gathered, and a detailed assessment is to be presented in another article. The average heavy metal concentrations in these minesoils are shown in Fig. 2. Soil pH ranged from 4.37 to 7.88 (averaged 6.21, 6.00, and 5.68 for Bayi, Lipu and Pingle, respectively), indicating an acid nature. Overall, Bayi minesoil had the highest Cd; Lipu had the highest Pb and Cr, and Pingle had the highest Zn, Mn and Cu (but no significant difference to the Cu level in Lipu). For Cd, with soil pollution warning threshold being 0.3 mg/kg according to the soil quality standard (GB15618-1995), these minesoils had substantially higher Cd levels (28 to about 94 times the warning value).

Table 6 presents the Bioaccumulation Factor (BAF) of the grown edible crops. None of the BAFs were beyond 1, and only six crops had BAFs for Cd larger than 0.5. Of the six metals studied, crops tended to have stronger Cd accumulation in edible parts; thus, Cd is the most important food safety threat. For other major contaminants, Pb and Cr, BAFs were largely below 0.1, and these may result from their relatively lower phyto-availability (2.03% for Pb, and 1.2% for Cr with 0.1M HCl extraction). Crops generally had very low Mn-accumulation ability (most BAFs were less than 0.05), but tea leaf in Bayi contained unusually high Mn contents, and this is in agreement with other results showing that tea tree is an Al and Mn accumulator.

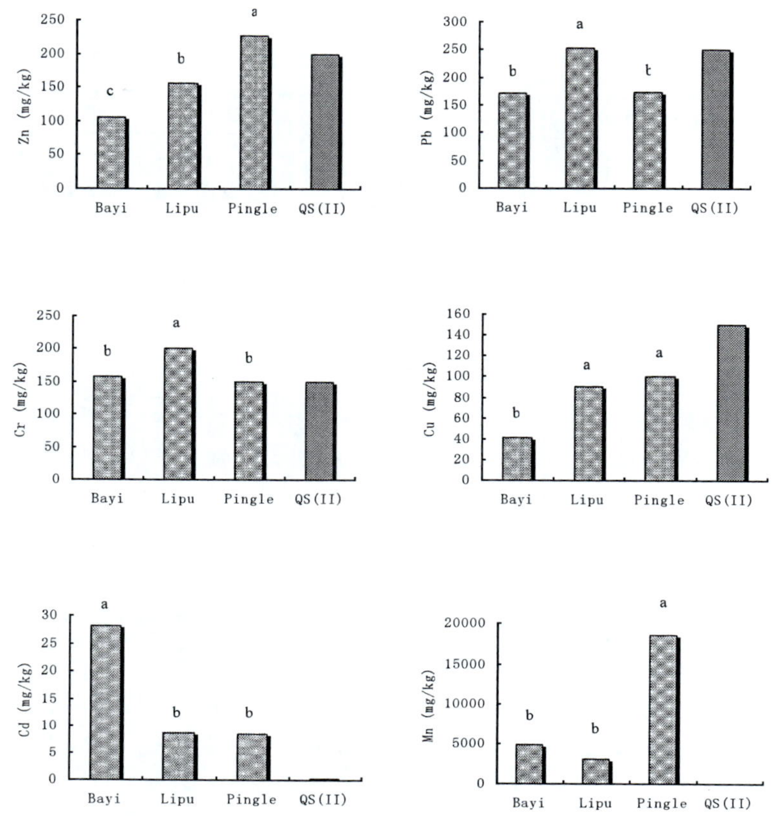

Figure 2. Heavy metal concentrations of soils from the three Mn mine wastelands. Different letters above bars indicate a significant difference ($P < 0.05$) using LSD test. QS(II) represents the soil quality standard value (GB15618-1995, Grade II for pH<6.5), indicating a pollution warning threshold. There was no standard value given for Mn.

DISCUSSION

Safety of Agronomic Crops Grown on the Reclaimed Mine Wastelands

Food safety problems in China can be traced largely to cultivated land. The substrate that crops grow is a fundamental warranty of agrarian product quality. Heavy metal contamination of staple foods has aroused a lot public concern not because of its actual harmful consequences (unlike pesticide residues) but because of the recent rejected exported food commodities and of publicity of heavy metal toxicities (Pb and Cd, in particular) by mass media. However, heavy metal pollution of mineland crops is far away to be addressed by the public as well as the local governments mainly because the pollution effect is usually slow and limited to regions. A recent study on a Dabaoshan mine area (a large multi-metal sulfide mine) of north Guangdong province revealed a horrible 'cancer village' – over 210 villagers of all ages died of cancer, 8.7 times of the national average (Chen, 2005), a tragic outburst of severe heavy metal contamination 30 years after the mining.

Table 6. Biological Accumulation Factors (BAFs) of grown crops on the three Mn mine wastelands in Guangxi

Crop	BAF of crops grown in mine wasteland regions					
	Zn	Pb	Cr	Cu	Cd	Mn
Bayi						
Peanut	0.330	0.046	0.015	0.250	0.030	0.006
Soybean	0.405	0.054	0.027	0.168	0.042	0.004
Sweet potato (tuber)	0.100	0.065	0.052	0.148	0.047	0.008
Sweet potato (leaf)	0.300	0.086	0.071	0.358	0.089	0.040
Cassava	0.143	0.034	0.031	0.119	0.066	0.002
Orange	0.096	0.038	0.017	0.093	0.062	0.002
Watermelon	0.347	0.055	0.021	0.117	0.065	0.010
Sugarcane	0.122	0.050	0.049	0.072	0.077	0.021
Tea	0.906	0.076	0.059	0.234	0.027	0.372
Lipu						
Peanut	0.230	0.040	0.027	0.074	0.567	0.014
Soybean	0.296	0.080	0.026	0.155	0.611	0.023
Sweet potato (tuber)	0.105	0.080	0.040	0.003	0.573	0.003
Sweet potato (leaf)	0.191	0.042	0.055	0.085	0.100	0.030
Cassava	0.146	0.089	0.036	0.081	0.643	0.002
Orange	0.079	0.135	0.032	0.012	0.176	0.003
Sugarcane	0.082	0.102	0.037	0.086	0.740	0.014
Nachi pear	0.114	0.022	0.085	0.079	0.015	0.002
Persimmon	0.080	0.098	0.052	0.008	0.354	0.011
Peach	0.126	0.041	0.036	0.056	0.049	0.003
Chinese Chestnut	0.106	0.076	0.025	0.066	0.383	0.038
Shallot	0.263	0.065	0.064	0.059	0.402	0.068
Garlic	0.212	0.066	0.044	0.063	0.548	0.048
Capsicum	0.218	0.125	0.094	0.118	0.418	0.011
Eggplant	0.109	0.037	0.077	0.069	0.042	0.063
Chinese radish	0.273	0.060	0.055	0.049	0.258	0.017
Cowpea	0.265	0.079	0.038	0.047	0.408	0.135
Endive	0.482	0.047	0.054	0.066	0.196	0.031
Lettuce	0.348	0.034	0.027	0.061	0.379	0.032
Pingle						
Peanut	0.143	0.198	0.094	0.058	0.186	0.001
Chinese Chestnut	0.080	0.040	0.046	0.106	0.234	0.008
Soybean	0.195	0.060	0.090	0.128	0.013	0.004
Cowpea	0.195	0.038	0.101	0.046	0.046	0.021
Sweet potato (tuber)	0.065	0.178	0.077	0.037	0.288	0.007
Cassava	0.081	0.184	0.024	0.016	0.248	0.003
Mandarin	0.045	0.236	0.103	0.023	0.397	0.002
Persimmon	0.038	0.239	0.031	0.013	0.315	0.005
Peach	0.072	0.041	0.068	0.034	0.018	0.001
Taro	0.297	0.137	0.089	0.041	0.306	0.030
Ginger	0.246	0.202	0.120	0.069	0.454	0.215
Tomato	0.086	0.048	0.122	0.069	0.059	0.001
Capsicum	0.105	0.133	0.079	0.094	0.333	0.007
Overall Average	0.198	0.087	0.056	0.086	0.250	0.032

In comparison with the many studies of food contamination around urban areas (Bai, 2004; Chen et al., 2004; Fu & Li, 1999; Li et al., 2000; Liu & Chen, 2004), safety of mineland crops was rarely investigated (Garcia et al., 1974; Garcia et al., 1979; Zhu et al., 2007) worldwide although these plantation practices were fairly common in developing countries especially those facing population pressure. In a field experiment of crop plantation on copper mine tailings mixed with loess (soil to tailings, 2:1 w/w), Zhou et al. (2002) reported that the As, Cd, and Pb concentrations in corn, sorghum, peanut, and soybean were lower than the prescribed allowable limits and thus were considered that they were safe to eat.

However, the conclusion is under question since other metals like Cr and Cu have exceeded the allowable standards, still posing health risk to consumers. In the corn grown on a Pb-Zn minesoil of Liaoning Province (North China), the Cd and Pb concentrations exceeded the standards by 1.5 and 2.0 times, respectively (Gu et al., 2005). Also in Baiyin Pb-Zn mine of northwest China, the Pb, Cd, Zn and Cu contents in forage grass and grain were greatly and significantly higher than those of uncontaminated sites, and grazing horses and sheep contained very high Pb and Cd levels in their blood, hair, liver, kidney, and skeleton, and showed obvious morbid symptoms in appearance (Liu, 2005).

In the present study, sugarcane (Bayi), tea (Bayi), Chinese chestnut (Lipu), and peach (Pingle) were all planted in large areas (>10 ha), and these products would enter human's food chain directly nearby and remotely through marketing and raw product processing. In terms of P_N, these four main crops were all severely contaminated, unacceptable as food according to the current food safety standards. Other crops were also heavily contaminated, but the health risk was mostly confined to the local inhabitants. Local governments should exercise monitoring programs of these crops and educate the mine-area residents to avoid the possible toxic effect in the long term.

Implications for Restoration of Mine Wastelands

The aim of restoration of mine wastelands is to remediate ecological destruction and reduce pollution dispersion. If rehabilitation of mine spoils causes another serious pollution to human being, the loss outweighs the gain. A general tendency of mineland rehabilitation in China is the utilitarian reclamation for agriculture. In fact, reclamation for planting crops is usually the last option of restoration in developed countries (Cook & Johnson, 2002) because it is very costly to meet the stringent soil requirement. The lax environmental controls and food safety enforcement in China as well as shortage of cultivable lands may account for this practice, and in Guangxi, this practice has been encouraged to some extent by the local government (Li, 2006; Li et al., 2007) due to severer lack of arable land. However, this restoration mode must be reconsidered carefully or modified.

First, China Environmental Quality Standard for Soils requires soils for agricultural crops meet Grade II criteria (see Fig. 2 for reference). Unfortunately almost all minesoils, especially of the metal-mined wasteland, can not satisfy the standard prior to remediation. Thus simple reclamation for crop growth is of high risk unless sufficient treatments, e.g., separation layer, cover of guest soil over 50-cm deep (Wong, 2003), or substantial substrate amendments (Li, 2006; Wong & Luo, 2003), are in place before planting. Even so, the yield of crops may reduce and quality be compromised. Another option is to choose low-accumulation cultivars of crops (Hu, 2004; Yao et al., 2006) or plant non-edible agronomic crops like ramee or use for pulpwood and charcoal wood.

Second, mono culture of crops is not good for restoration of pre-mining biodiversity. Biodiversity, vegetation structure, and ecological processes are the most important ecosystem attributes to evaluate restoration success (Ruiz-Jaen & Aide, 2005). In terms of these attributes, few of the China's reclamation efforts for agricultural plantation can be deemed successful although some may help generate short-term income for local residents.

Finally, miners and restorers need to leap out of this utilitarianism-oriented reclamation. Restoration of minelands can have much more diverse functions, such as nursery, forestry, biodiversity conservation, recreation and tourism, providing habitat to wildlife, checking soil and wind erosion, or just beautifying the damaged landscape.

ACKNOWLEDGMENTS

This work was jointly supported by the National Science Foundation of China (Grant No. 30560032) and Guangxi Science Foundation (Grant No. Guikeji 0575047). Also, we thank Guangxi Normal University for financial assistance with a Startup Research Grant for the Introduced Talents. Appreciation was attributed to the two anonymous scientists who examined this manuscript in detail and made it more readable. Prof. Yinian Zhu at Guilin University of Technology reviewed the article and gave constructive improvement suggestions. In addition, Ms Shengxiang Yang helped with the field work and Mr. Chunqiang Chen assisted with the laboratory test.

REFERENCES

Bai, H. J. (2004). Analysis on content of Pb, Cr and Cd in vegetables from the market of Taiyuan city and their safety evaluation. *China Safety Science Journal, 12,* 78-81.

Chan, X. X. & Shi, X. D. (2001). Heavy metal pollution of soils and food safety. *Yunnan Environmental Science, 20 (suppl.)*, 21-24.

Chen, G. F., Huang, W. J., Zhang, L. M., Huang, K. F. (2004). Survey and evaluation of heavy metal contamination of soils and vegetables in Nanning. *Guangxi Agricultural Science*, *35(5)*, 389-392.

Chen, H. M. (2006). *Environmental Pedology*. Beijing: Science Press.

Chen, N. C. (2005). Environmental problems of Dabaoshan Mine Area and their potential harm. *Program and Abstract.* Conference on China's Soil Environmental Problems and Strategies, Guangzhou, November 13-16, pp. 119.

Cooke, J. A. & Johnson, M. S. (2002). Ecological restoration of land with particular reference to the mining of metals and industrial minerals: a review of theory and practice. *Environmental Review, 10*, 41-71.

Fu, Y. H. & Li, Y.J. (1999). Investigation on pollution of vegetables in Shenyang suburban areas. *Agro-environmental Protection*, *18(1)*, 36-37.

Garcia, W. J., Blessin, C. W., Inglett, G. E., Carlson, R. O. (1974). Physical-chemical characteristics and heavy metal content of corn grown on sludge-treated strip-mine soil. *Journal of Agricultural and Food Chemistry, 22(5)*, 810-815.

Garcia, W. J., Blessin, C. W., Sandford, H. W., Inglett, G. E. (1979). Translocation and accumulation of seven heavy metals in tissues of corn plants grown on sludge-treated strip-mined soil. *Journal of Agricultural and Food Chemistry*, *27(5)*, 1088-1094.

Gu, J. G., Lin, Q. Q., Hu, R., Zhuge, Y. P., Zhou, Q. X. (2005). Translocation behavior of heavy metals in soil-plant system - A case study from Qingchengzi Lead-Zinc Mine in Liaoning Province. *Journal of Agro-Environment Science, 24(4)*, 634-637.

Hu, P. S. (2004). Soil Cd toxicity and screen and modification of low-accumulation rice cultivars. *China Rice*, *2*, 10-12.

Huayi net - European Chinese news, *http://news.chinesewings.com/cgi-bin/site/c.cgi*=286 *id=20050326180665191*. Seen on March 6, 2007.

Li, M. S. (2005). Ecological Restoration of the Metal-Mined Wastelands in Guangxi. *Program and Abstract*, Conference on China's Soil Environmental Problems and Strategies, Guangzhou, November 13-16 pp. 91.

Li, M. S. (2006). Ecological restoration of mineland with particular reference to the metalliferous mine wasteland in China: a review of research and practice. *Science of the Total Environment, 357*, 38-53.

Li, M. S., Luo, Y. P., Su, Z. Y. (2007). Heavy metal concentrations in soils and plant accumulation in a restored manganese mineland in Guangxi, South China. *Environmental Pollution, 147*, 168-175.

Li, Q. L., Zhao, Z. J., Huang, Y. (2000). Present situation of quality in soils and vegetables of the vegetable bases in suburbs of Chongqing. *Chongqing Environmental Science, 22(6)*, 33-36.

Liu, G. H. & Shu, H. L. (2003). Research progress of ecological restoration in mine spoils. *Jiangxi Forestry Science and Technology, 2,* 21-25.

Liu, J. H. & Chen, Y. C. (2004). Primary investigation on the contamination pattern of vegetables by heavy metals in Chinese cities. *Studies of Trace Elements and Health, 21(5)*, 42-44.

Liu, Z. P. (2005). The bioactivity of environment heavy metal pollutants in the vicinity of non-ferrous metal smelters. *Acta Ecologica Sinica, 25(2)*, 273-278.

McLaughlin, M. J., Parker, D. R., Clarke, J. M. (1999). Metals and micronutrients - food safety issues. *Field Crops Research, 60*, 143-163.

Ruiz-Jaen, M. C. & Aide, T. M. (2005). Restoration success: How is it being measured? *Restoration Ecology, 3*, 569-577.

Singh, A. N., Zeng, D. H., Chen, F. S. (2005). Heavy metal concentrations in redeveloping soil of mine spoil under plantations of certain native woody species in dry tropical environment, India. *Journal of Environment Sciences, 1*, 168-174.

Sun, T. X., Li, P. J., Zhou, Q. X. (2005). *Mechanisms of soil pollution and remediation technologies*. Beijing: Science Press.

Wong, M. H. (2003). Ecological restoration of mine degraded soils with emphasis on metal contaminated soils. *Chemospere, 50*, 775-780.

Wong, M. H. & Luo, Y. M. (2003). Land remediation and ecological restoration of mined land. *Acta Pedologica Sinica, 2*, 161-169.

Yao, H. M., Du, T. T., Su, D. C. (2006). Cadmium uptake and accumulation in vegetable species in *Brassica cruciferae*. *Chinese*, *22*, 291-294.

Yang, S. X., Li, M. S., Lai, Y. P., Luo, Y. P. (2007). Dominant plants and their heavy metal concentrations in Mn mine wastelands, Guangxi. *Journal of Guangxi Normal University (Natural Science Edition)*, *25(1)*, 108-112.

Zhong, W. K., Fan, Y. B., Wang, M. J. (2001). Pollution of heavy metals on crops and its countermeasures in China. *Agro-environmental Protection, 20(4)*, 270-272.

Zhou, L. B., Dai, H. W., Wu, Y. J. (2002). Crop plantation on the copper mine tailings ponds in Hujiayu. *Mining Technology, 2(2)*, 54-56.

Zhu, Y., Yu, H., Wang, J., Fang, W., Yuan, J., Yang, Z., 2007. Heavy metal accumulations of Asparagus bean cultivars grown in soil contaminated with Cd alone and with multiple metals (Cd, Pb and Zn). *Journal of Agricultural and Food Chemistry, 55(3)*, 1045-1052.

Reviewed by Prof. Dr. Yinian Zhu

Department of Resources and Environmental Engineering, Guilin University of Technology, Guilin, CHINA. Tel: 86 773 5897016; Fax: 86 773 5897405; email: zhuyinian@glite.edu.cn.

INDEX

A

B

C

D

E

F

N

O

P

Q

R